"Scientist of the Islamic Era" is a book volumes; the present book is volume 2 titled "Medical Scientists of the Islamic Era". It covers 36 medical scientists encompassing physicians, nurses, surgeons, herbalists, medical researchers, and medical writers. They commanded exceptional breadth in their learning and deepest insights in their specializations; they greatly strengthened the foundations and expanded the frontiers of the fields of knowledge in the Medical Sciences.

Each scientist is briefly described. First, the name of the scientist is disambiguated, and an attempt is made to correct the misrepresentations common in the European translations. Salient scientific contributions of each scientist are briefly highlighted, a difficult task because of the fact that many of these scientists were polymaths. For each scientist we have provided a biographical summary to help picture their love and craving for knowledge, and the motivations and opportunities for them to do their research. The period is Part 1 of the Islamic Era, from 610 to 1400 AD.

It is our objective that this second volume in the series will inform the Muslims about the wealth of their scientific heritage, and the next generations will feel inspired to surpass the excellence of their ancestors to enrich their heritage further, and be, like their ancestors, the flag bearers of world civilization in the medical sciences.

Muslims are now excelling in scientific and technology research with superb agility, and this series on Scientists of the Islamic Era will further stimulate this Renaissance in the Muslim world.

Medical Scientists of the Islamic Era

Volume 2 of the 8-volume series on

Scientists of the Islamic Era

Medical Scientists of the Islamic Era

Volume 2 of the 8-volume series on

Scientists of the Islamic Era

Abdur Rahim Choudhary, Ph.D.

Muslim Voice

MV Publishers

Published by MV Publishers, a subsidiary of Muslim Voice,
12719 Hillmeade Station Dr, Bowie, MD 20720, USA.
MVPublishers@muslimvoice.org

ISBN 978-1-956601-14-5

First edition 2023
United States of America

Choudhary, Abdur Rahim, 1944–
8-Volume Series on Scientists of the Islamic Era,
Volume 2, Medical Scientists of the Islamic Era
Muslim Voice

ISBN 978-1-956601-14-5

To the Muslim Ummah

Content

Preface to the 8-volume series on Scientists of the Islamic Era

For a period of more than a millennium, Muslim Scientists have done foundational research in all scientific disciplines, and also greatly expanded the frontiers of science. However, our people often do not have a clear idea about our scientific heritage. We decided to write a series of books on "*Scientists of the Islamic Era*" that would be readily available to our generation and the coming generations, and provide motivation for excellence in the world civilizational dialogue, as well as to know our religious inspiration for scientific research and progression.

The young generations, especially those in Europe and Americas, have now opened their hearts and minds with a renewed desire for the truth about Islam and Muslims, being less influenced by historical biases and religious prejudices. The eight books in the series on *Scientists of the Islamic Era* seek to serve their youthful thirst for the truth.

Another reason for this series on "*Scientists of the Islamic Era*" is to produce a consciousness among the present-day academicians and scientists about the foundational contributions that the Muslim scientists made to all scientific disciplines, as well as how they expanded the frontiers of these disciplines. This fact is evidenced in the books in this series. However, this fact is not widely known because the present-day literature does not reference these original sources. The chain of scholarly references ends in European Renaissance, with occasional references to Greek scientists, but bypassing the millennium worth of research by Muslim scientists, who established the foundational principles and greatly expanded the frontiers of science.

In addition, the work seeks to fill a void, as no such series of books currently exists.

Islamic Era constitutes the period from 610 AD, when the Prophet received his first revelation, to 1922 AD, when the Ottoman Caliphate ended and the Turkish Republic began. We have divided the period in two parts: part 1 from 610 to 1400, and part 2 from 1400 to1922. The era is divided at an epoch when much of the works by the Muslim Scientists had already been translated into European languages, had become widely available, and had begun to produce Renaissance in Europe.

Each of the two parts of the Islamic Era is covered by the following four volumes, eight volumes in all.

1. Volume 1 is for Natural Sciences that include mathematics, astronomy, cryptoanalysis, chemistry, cartography, physics, and engineering based on these disciplines such as mechanics, automation, and robotics.

2. Volume 2 is for the Medical Sciences that include physicians, nurses, surgeons, herbalists, medical researchers, and medical writers.

3. Volume 3 is for the Social Sciences that include philosophers, historians, physical geographers, qadhis, and hadith narrators, as well as the conventional sociology, political science, management sciences, economics, business, trade, anthropology, and linguists.

4. Volume 4 is for the Religious Sciences that include analogists, mohaddasin (historical fact checkers), jurisprudents, mofassarin (Quranic exegetists), and spiritualists (sciences of the tariqas).

We present this series of books to the readers to share with them the wealth of scientific excellence that these scientists contributed to the world civilization and in producing Renaissance in Europe; to bring awareness to the Muslim readers about their role as the torch bearers of science and civilization; to serve the upwelling thirst that the young generation have for the truth about Islamic civilization; and to urge the academicians and researchers of the world, especially the Europeans and Americans, to learn and celebrate the Muslim giants of science upon whose shoulders they stand, and without whom the present-day scientific achievements could not have been possible.

Researchers like Professor Fuat Sezgin have devoted their lives to research and investigate contributions of the Muslim scientists. He has edited 1600 volumes. Such work is invaluable for projects like ours. His works are summarized online in Wikipedia articles, and are available under GNU free document license. We have used such sources, but performed extensive recension, critical editing and reorganization.

We aim to serve our community, inspire them and our coming generations, and inform them of their role as torch bearers of excellence in world science, technology, and civilization. Acknowledgment is also due to Professor Abdur Rahim Choudhary and Ms. Yasmeen Sultana Choudhary whose total dedication made the work possible.

The Muslim scientists lived an integrated life with no conflict between the religion and the scientific passions; and, unlike the many present-day scientists, a question never occurred that their scientific passion somehow needed to be separate from their religious inspirations. This is also obvious from the fact that most scientists were themselves experts in Islamic jurisprudence, hadith and Quran. In reality, their scientific work was also their religious worship because Islam showed them the necessity to do science, provided the motivation for it, and supported their scientific passion by equating it with religious worship. No wonder they achieved scientific excellence with amazing integrity, generosity and grace.

The Muslim education system was very different from what the world has today; and judging from the results, it was greatly more affective and better integrated into overall life. The biographical summaries will provide tiny glimpses into this system. It avoided narrow specializations and produced polymaths. We have Muslim scientists who are simultaneously excellent Quranic scholars, jurisprudents, mathematicians, astronomers, physicists, medical professionals, chemists, botanists, physiologists, poets, men of letters, grammarians, etc. To keep this series on "Scientists of the Islamic Era" brief we have included each scientist only once, under a category we deemed appropriate, given their biography. As a consequence, for instance, not all natural scientists appear in volume 1 which is dedicated to the Natural Scientists, because we decided to include some in Volume 2 which is dedicated to the Medical Scientists; similarly, not all medical scientists appear in Volume 2 because we decided to include some in Volume 1. This situation invariably occurs across the board. Another consequence is that, for instance, if a scientist is included in Volume 2 for the Medical Scientists, his or her achievements in Astronomy, for instance, are not highlighted.

For such reasons, the preface to the series will get edited as the series progresses and situations for it arise.

The scientists are listed in chronological order, allowing an opportunity to correlate scientific tides and ebbs with political and religious ups and downs.

They could have been ordered according to the significance of their scientific contributions; that, however, is problematic because it is difficult, if not impossible, to assess the importance of research and compare across different scientific disciplines within sciences.

The order could have been sequenced according to how well the scientists are known today; that too is problematic because not all excellent scientists are well-known today, and, those who are, generally are made famous by the European commentators, who often did not know their works in original Arabic, and did not reflect the actual significance of their research. The well-known-ness is fairly arbitrary. For instance, Omar Khayyam is celebrated today for his Rubaiyat, which was something he did on the side, while his real works were in mathematics, a fact that is largely obscured.

This series of books should add to the impulse that is now thrusting the Muslims into the world of science and technology with increasing excellence in their achievements, signaling that their own Renaissance has now begun.

Muslim Voice
Bowie, MD, USA.
July 29th, 2022.

Preface to the First Edition of Volume 2: Medical Scientists

"Scientist of the Islamic Era" is a book series encompassing eight volumes. The present book is volume 2 titled "Medical Scientists of the Islamic Era" that covers 36 physicians, nurses, surgeons, herbalists, medical researchers, and medical writers. The period of coverage is part 1 of the Islamic Era, from AD 610 to 1400.

Most medical scientists in this book are multidisciplinary and interdisciplinary; and they also excelled in jurisprudence, hadith, philology, and poetry. They commanded exceptional breadth in their learning and deepest insights in their specializations; and, thus, greatly strengthened the foundations and expanded the frontiers of fields of knowledge.

It is our objective that this second volume in the series will inform the Muslims about the wealth of their scientific heritage, and the next generations will feel inspired to surpass the excellence of their ancestors to enrich their heritage further, and be, like their ancestors, the flag bearers of world civilization regarding the medical sciences. Our objective is also for the academic community to learn the truth about how science grew by leaps and bounds during the Islamic era. The book series shall quench the thirst of the youth, especially in Europe and Americas, to discover the truth about Muslim contributions to the world science, technology, and civilization.

Muslims are now excelling in scientific and technology research with superb agility; the books in the series on "Scientists of the Islamic Era" are expected to add impetus to this Renaissance in the Muslim world.

Abdur Rahim Choudhary, Ph.D.
Bowie, Maryland, USA
arc@muslimvoice.org
July 29th, 2022.

Medical Scientists

Medical Sciences community in Islamic Era was dominated by the Muslim scientists; the European scientists during this time were virtually nonexistent, owing to Europe being in the "Dark Age". When they started to emerge a little before the European Renaissance, they did so based on the research works of the Muslim scientists done for the prior seven centuries, which had already been translated into European languages, and had become broadly available.

These facts are obvious even if one examines not the entire scientific works by the Muslim scientists but only a subset of those that had been very visibly translated into European languages.

This book describes 36 medical scientists from part 1 (610-1400) of the Islamic era covering the disciplines of physicians, nurses, surgeons, herbalists, medical researchers, and medical writers. Each scientist is briefly described. First, the name of the scientist is disambiguated, and an attempt is made to correct the misrepresentations all too common in the European translations. Salient scientific contributions of each scientist are briefly highlighted, a difficult task because of the fact that most of these scientists were polymaths. For each scientist we have provided a biographical summary to help picture their love and craving for knowledge and the motivations and opportunities for them to do their research.

The list of 36 medical scientists, that are covered in this edition of the book, is given in the table below, in chronological order. Each entry in the table includes the year of death and a one-line description, including the name of the scientist, the time period in parenthesis, and the area(s) of specialization within the medical sciences.

Table of 36 Medical Scientists covered in this edition of the book.

620	Rufaida Al-Aslamia (b. 620), physician
634	Nafi ibn al-Harith (d. 13 AH/634–35), physician
650	Ibn Abi Ramtha (7th century), physician
650	Al-Shifa' bint Abdullah (7th century), practiced herbal medicine
650	Ibn Uthal (7th century), physician
690	Paul of Aegina (625-690), Physician
850	Al-Ruhawi (9th century), physician
873	Hunayn ibn Ishaq al-Ibadi (808-873), translator and physician.
910	Abū Yaʿqūb Isḥāq ibn Ḥunayn (830-910), physician and traveler.
935	Abu Bakr Al-Razi (865-935) Physician, Philosopher, Chemist
950	Ibn al-Jazzar (10th century), influential 10th-century physician and author
950	Al-Tamimi (10th-century), physician from Palestine
976	Al-Jabali (d. 976), physician and mathematician from Al-Andalus
990	Ammar al-Mawsili (10th century, b. Mosul), ophthalmologist and physician
994	Ibn Juljul (c. 944–c. 994), physician and pharmacologist
1010	Ali ibn Isa al-Kahhal (fl. 1010), physician and ophthalmologist
1013	Al-Zahrawi (936–1013), Father of surgery, wrote *Al-Tasrif*, a thirty-volume collection of medical texts, upon which European surgical procedures are based.

1029	Ibn al-Kattani (951–1029), physician, scholar, philosopher, astrologer, man of letters, and poet
1033	Ibn al-Thahabi (d. 1033), physician and author of the first known encyclopedia of medicine
1037	Ibn Sina (Avicenna) (987-1037), Polymath, Physician, Astronomer, Philosopher
1050	Ibn Jazla (11th century), physician and author of influential treatise on regimen
1074	Ibn al-Wafid (997–1074), pharmacologist and physician
1075	Ibn Butlan (1038, Baghdad – 1075), physician
1161	Ibn Zuhr (1091–1161), prominent physician
1165	Al-Ghafiqi (d. 1165), 12th-century oculist
1174	Abu al-Majd ibn Abi al-Hakam (d. 1174), physician, musician
1213	Ibn Hubal (1122–1213), physician, scientist and author of a medical compendium
1230	Al-Dakhwar (1170, Damascus – 1230), physician
1231	Abd al-Laṭīf al-Baghdādī (1162-1231), physician, philosopher, historian, Arabic grammarian and traveler
1241	Rashidun al-Suri (1177–1241), physician and botanist
1248	Ibn al-Baitar (1197, Malaga – 1248, Damascus), pharmacist, botanist, physician
1270	Ibn Abi Usaybi'a (1203-1270), physician from Syria
1286	Ibn al-Quff (1233–1286), physician
1288	Ibn al-Nafis (1213–1288), physician and author, the first to describe pulmonary circulation, compiled a medical encyclopedia and wrote numerous other works
1292	Al-Suwaydi (1204–1292), physician
1348	Ibn al-Akfani (1286, Sinjar – 1348, Cairo), encyclopedist and physician

We expect that this list will be expanded in subsequent editions, as further research is carried out.

A brief description for each scientist is provided, each in a separate subchapter. The 36 subchapters, that follow, are each dedicated to a single medical scientist. Some chapters are short, while others are detailed. Information on these topics is not abundant because the existing research is at best sporadic, and is mostly championed by individuals or small groups. There is a strong need for more detailed studies, on sustained and institutional bases, on an expansive scale.

The present series of eight volumes is offered in this context. They are intended for the educational and research institutes, at national and international levels, to provide encouragement for further work focused along these lines.

1.   Rufaida Al-Aslamia

Rufaida Al-Aslamia (also transliterated Rufaida Al-Aslamiya or Rufaydah bint Sa`ad)

(Arabic: رفيدة الأسلمية),

(born approx. 620 AD),

was medical professional who invented mobile medical services.

Scientific Contributions

Rufaida Al-Aslamia implemented her clinical skills and medical experience into developing the first-ever documented mobile care units that were able to meet the medical needs of the community. The scope of the majority of her work in her organized medical command units consisted primarily in hygiene and stabilizing patients before further and more invasive medical procedures.

During military expeditions, Rufaida Al-Aslamia led groups of volunteer nurses who went to the battlefield and treated the casualties. She participated in the battles of Khandaq, Khaibar, and others.

During times of peace, Rufaida Al-Aslamia continued her involvement with humanitarian efforts.

Rufaidah had trained a group of women companions as nurses. When the Prophet's army was getting ready to go to the battle of Khaibar, Rufaidah and the group of volunteer nurses went to the Prophet. They asked him for permission "O Messenger of Allah, we want to go out with you to the battle and treat the injured and help Muslims as much as we can". He permitted them to go. The nurse volunteers did a superb job, and the Prophet assigned a share of the bounty to Rufaidah. Her share was equivalent to that of soldiers who had fought.

This was the first ever official recognition, in the world, for the prime significance of the medical services during war time.

Each year the Royal College of Surgeons in Ireland at the University of Bahrain awards one student the prestigious Rufaida Al-Aslamia Prize.

Biographical Summary

Rufaida Al-Aslamia was born in 620 AD. Among the first people in Madina to accept Islam, Rufaida Al-Aslamia was born into the Bani Aslem tribe of the Kazraj tribal confederation in Madina, and gained fame for her contribution with other Ansar women who welcomed the Islamic prophet, Muhammad, on arrival in Madina.

Rufaida Al-Aslamia is depicted as a kind, empathetic medicine professional and a good organizer. With her clinical skills, she trained other women, including the famous female companion of Muhammad, Ayesha, to be nurses and to work in the area of health care.

She also worked as a social worker, helping to solve social problems associated with the disease. In addition, she assisted children in need and took in orphans, and helped the poor.

Born into a family with strong ties to the medical community, Rufaida's father, Sa`ad Al Aslamy, was a physician and mentor under whom Rufaida initially obtained clinical expertise. Devoting herself to medical profession and taking care of sick people, Rufaida Al-Aslamia became an expert healer. Although not given responsibilities held solely by men such as surgeries and amputations, Rufaida Al-Aslamia practiced her skills in field hospitals in her tent during many battles as Muhammad used to order all casualties to be carried to her tent so that she might treat them with medical expertise.

2. Nafi ibn al-Harith

Nāfiʿ ibn Al-Ḥārith ibn Kaladah ath-Thaqafi

(Arabic: نَافِع ابْن الْحَارِث ابْن كَلَدة الثَّقَفِي),

(died 634 – 635 CE),

was a physician from the Banu Thaqif.

Scientific Contributions

He treated Saʿd ibn Abi Waqqas and Abu Bakr. When Abu Bakr was dying, Nāfiʿ ibn Al-Ḥārith designated his illness as poisoning.

Nāfiʿ ibn Al-Ḥārith was trained in Yemen. He wrote a book: Dialog in Medicine.

He was a teacher at the Academy of Gundishapur in Persia.

Biographical Summary

Nāfiʿ ibn Al-Ḥārith died in 634-35 AD. He was a half-brother of Nufay ibn al-Harith (also known as Abu Bakra bin Kalada al-Thaqafi at-Thaifi).

3.   Ibn Abi Ramtha

Ibn Abi Ramtha al-Tamimi

from Banu Tamim tribe,

(Arabic: ابن أبي رمثة التميمي),

was a physician. who lived during the lifetime of Prophet Muhammad.

Scientific Contributions

He was a skilled practitioner, and occasionally also practiced surgery.

Biographical Summary

Ibn Abi Ramtha was a companion of the Prophet from Banu Tamim tribe in Medina.

4.   Al-Shifa' bint Abdullah

Al-Shifaa bint Abdullah

(Arabic: الشفاء بنت عبد الله),

whose given name was Layla,

was a healer who practiced herbal medicine; she was a companion of Prophet Mohammad.

Scientific Contributions

Al-Shifaa bint Abdullah was a healer, as her name "Al-Shifa" suggests. She practiced herbal medicine.

She was a wise woman, as suggested by the fact that the Prophet consulted with her on best practices.

Biographical Summary

She was the daughter of Abdullah ibn Abdshams and Fatima bint Wahb and a member of the Adi tribe of the Quraysh in Mecca. She married Abu Hathma ibn Hudhayfa, and they had two sons, Sulayman and Masruq.

She had a reputation as a wise woman. Her by-name Al-Shifaa means "the Healer" and indicates that she practiced medicine. At a time when barely twenty people in Mecca could read and write, Al-Shifaa was the first woman to acquire medical expertise. She taught calligraphy to many, including, her relative, Hafsa bint Umar, a wife of the Prophet. The two women remained friends.

11

Al-Shifaa became a Muslim in Mecca and was among the first to join the emigration to Medina. There she had a house between the mosque and the market. The Prophet sometimes consulted her about best practices.

5.   Ibn Uthal

Ibn Uthal or Ibn Athal

(Arabic: ابن أثال),

served as a personal physician of the caliph Mu'awiya I.

Scientific Contributions

Ibn Uthal was skilled in toxicology. He served as a physician of the first Umayyad caliph.

Biographical Summary

Ibn Uthal was killed in the 7th century in a revenge attack.

6.   Paul from Aegina

Paul from Aegina

(625-690),

was a medical writer.

Scientific Contributions

Paul from Aegina was presumably a physician, as he is reported to have been consulted by midwives for childbirth. Soudas compiled a reportedly 30000 entry collection on ancient Mediterranean world, and this collection mentions Paul. It is reported that Paul himself compiled books on ancient European medicines. He presumably was not aware of relatively recent works of Rufaida Al Aslamia and Ibn Abi Ramtha and others who were taking the field of medicine beyond what Europe knew. The books compiled by Paul are therefore likely to be relatively less useful. However, it seems that the Europeans merrily kept using these books until the more advanced research by the Muslim medical scientists were translated in to the European languages. Such translations ignited a Renaissance in Europe.

Biographical Summary

Nothing is known about Paul. As his name seems to suggest, he probably came from the island of Aegina. He seems to have visited Alexandria in Egypt and some people speculate that he was a traveler. The exact time when he lived is not known, but the Heresiographer

Ibn Jawzi mentions Paul and places him towards the second part of seventh century. He is reported to have died around 690 AD.

7. Al-Ruhawi

Ishāq bin Ali al-Rohawi

(Arabic: إسحاق بن علي الرهاوي),

was a 9th-century author of the first medical ethics book.

Scientific Contributions

His Ethics of the Physician contains the first documented description of a peer review process, where the notes of a practicing physician were reviewed by peers and the physician could face a lawsuit from a maltreated patient if the reviews were negative.

Al-Rohawi's most celebrated work is: Adab al-Tabib ("Practical Ethics of the Physician" or "Practical Medical Deontology"). The work consisted of twenty chapters on various topics related to medical ethics. This is the earliest work on medical ethics.

Al-Rohawi regarded physicians as "guardians of souls and bodies".

He also wrote the following books:

- Introduction to Dialectics for Beginners
- On Examination of Physicians
- He wrote two additional books; and compilation of additional works.

Biographical Summary

Al-Rohawi lived during the 9th century. He was probably from Al-Ruha, modern-day Şanlıurfa in Turkey, close to the border with Syria, which is often simply known as Urfa. Not much is known about Al-Rahawi.

Al-Rahawi begins his book with the words "In the name of Allah", the universal style of Muslim writers. This and additional body of evidence debunks some Western writers who state, without evidence, that Al-Rahawi was Christian and some make him Jewish.

8.   Hunayn ibn Ishaq al-Ibadi

ʾAbū Zayd Ḥunayn ibn ʾIsḥāq al-ʿIbādī

   (also Hunain or Hunein),

   (Arabic: أبو زيد حنين بن إسحاق العبادي),

   (808–873),

   was a translator and physician.

Scientific Contributions

Hunayn ibn Ishaq was a translator at the institution of House of Wisdom in Baghdad. He knew Greek, Persian, Arabic, and Syriac, and was able to translate compositions on philosophy, astronomy, mathematics, medicine, and even in subjects such as magic and oneiromancy.

Nonetheless, none of his extant translations credit the House of Wisdom, which questions the legitimacy of the scholarly view propagated by the Orientalists that the House of Wisdom was a glorified description for a translation project. It puts the assumption and stance in serious doubt because in actuality the House of Wisdom presumably might have been different from the translation project.

Some of Hunayn's most notable translations were his rendering of "De materia Medica", a pharmaceutical handbook. Another translation selection is "Questions on Medicine", a guide for novice physicians. Many R. Duval's published works on chemistry are themselves translations of Hunayn's translated renderings in Arabic.

In his efforts to translate Greek material, Hunayn ibn Ishaq was accompanied by his son Ishaq ibn Hunayn and his nephew Hubaysh. Hunayn. They would translate Greek into Syriac, and then he would have his nephew finish by translating the text from Syriac to Arabic. Finally, he would then seek to correct any of his partners' mistakes or inaccuracies he might find. This detail questions the claim of the orientalists regarding Hunayn having command over Arabic.

The translation process was not what the modern reader might imagine. It largely did not follow the text's exact lexicon. Instead, it would try to summarize the topics of the original texts and then in a new manuscript paraphrase it in Syriac or Arabic. It edited and redacted the available texts of technical works being translated, by comparing the information included therein with other works on similar subjects. Thus, these renditions were not simple translation; rather, they were interpretations of texts after researching the topics over which they range.

Hunayn says:

Galen's works were translated before me by a certain Bin Sahda. When I was young, I translated them from a faulty Greek manuscript. Later when I was forty, my pupil Hubaish asked me to correct the translation. Meanwhile a number of manuscripts had accumulated in my possession. I collated these manuscripts and produced a single correct copy. Next, I collated the Syriac text with it and corrected it. I am in the habit of doing this with everything I translate.

Hunayn wrote some works of his own in addition to the translation. In "How to Grasp Religion", Hunayn explains the truths of religion that include miracles. He also wrote "The Rules of Inflexion According to the System of the Greeks". In ophthalmology he wrote, "Book of the Ten Treatises of the Eye". It is probable that these were compilations rather than original research; for example, his medical works were pulled from Greek sources such as, Fi Awja al-Ma'idah (On Stomach Ailments) and al-Masail fi'l-Tibb li'l-Muta'allimin (Questions on Medicine for Students).

Hunayn was a respected physician and was appointed one of the personal physicians to Caliph al-Mutawakkil. Despite their relationship, the caliph became distrustful of Hunayn and imprisoned his physician for a year.

Biographical Summary

Hunayn was originally from al-Hirah, but he spent his working life in Baghdad, the center of learning. His father was a pharmacist, and he went to Baghdad in order to study medicine under renowned physician Yuhanna ibn Masawayh; however, he was expelled from school of medicine. Therefore, Hunayn left to study the Greek language in Greece. But the center of learning and knowledge and prosperity was the Muslim world. On his return to Baghdad, he sought employment as a translator.

Because of his connection with al-Hirah, some orientalists extrapolate to assume that Hunayn was a Christian.

9.   Ishaq Ibn Hunayn

Abū Yaʿqūb Isḥāq ibn Ḥunayn

(Arabic: إسحاق بن حنين),

(c. 830 Baghdad, – c. 910-1).

was a physician and translator, known for his biography of physicians.

Scientific Contributions

Ishaq ibn Hunayn is known for writing the biography of physicians. He is also known for his translations of Euclid's Elements and Ptolemy's Almagest.

Biographical Summary

Ishaq ibn Hunayn was the son of the translator Hunayn Ibn Ishaq (please pay attention to the Arabic names). He was born in Baghdad in 830 AD and died in 910.

10.   Abu Bakr Al-Razi

Abū Bakr Muḥammad ibn Zakariyyāʾ al-Rāzī

(Arabic: أبو بكر محمد بن زكرياء الرازي)

(c. 864 or 865–925 or 935 CE),

was a Persian physician, philosopher and alchemist; he is among the top most few figures in the history of medicine.

Scientific Contributions

A comprehensive thinker, al-Razi made fundamental and enduring contributions to medical science and various other fields, which he recorded in over 200 manuscripts, and is particularly remembered for numerous advances in medicine through his observations and discoveries. An early proponent of experimental medicine, he became a successful doctor, and served as chief physician of Baghdad and Ray hospitals. As a teacher of medicine, he attracted students of all backgrounds and interests. He was compassionate and devoted to the service of his patients, whether rich or poor. He was the first to clinically distinguish between smallpox and measles, and suggest sound treatment for the former.

Through translation, his medical works and ideas became known among European practitioners and profoundly influenced medical education in the West. Some volumes of his work Al-Mansuri, namely "On Surgery" and "A General Book on Therapy", became part of the medical curriculum in Western universities. He was probably the greatest and most original of all the physicians, and one of the most prolific as an author. Additionally, he has been described as the father of pediatrics, and a pioneer of obstetrics and ophthalmology. Notably, he became the first physician to recognize the reaction of the eye's pupil to light.

Examples of Al-Razi's Research Approach

- Psychology and psychotherapy:

Al-Razi was one of the world's first great medical experts. He is considered the father of psychology and psychotherapy.

- Smallpox vs. measles: On these diseases Al-Razi wrote:

Smallpox appears when blood "boils" and is infected, resulting in vapors being expelled. Thus, juvenile blood (which looks like wet extracts appearing on the skin) is being transformed into richer blood, having the color of mature wine. At this stage, smallpox shows up essentially as "bubbles found in wine" (as blisters). This disease can also occur at other times (meaning: not only during childhood). The best thing to do during this first stage is to keep away from it, otherwise this disease might turn into an epidemic.

Encyclopedia Britannica (1911) states: "The most trustworthy statements as to the early existence of the disease are found in an account by the 9th-century Persian physician Rhazes, by whom its symptoms were clearly described, its pathology explained by a humoral or fermentation theory, and directions given for its treatment."

Al-Razi's book "al-Judari wa al-Hasbah" (On Smallpox and Measles) was the first book describing smallpox and measles as distinct diseases. It was translated more than a dozen times into Latin and other European languages. Its lack of dogmatism and its reliance on clinical observation show al-Razi's medical methods. For example, he wrote:

The eruption of smallpox is preceded by a continued fever, pain in the back, itching in the nose and nightmares during sleep. These are the more acute symptoms of its approach together with a noticeable pain in the back accompanied by fever and an itching felt by the patient all over his body. A swelling of the face appears, which comes and goes, and one notices an overall inflammatory color noticeable as a strong redness on both cheeks and around both eyes. One experiences a heaviness of the whole body and great restlessness, which expresses itself as a lot of stretching and yawning. There is a pain in the throat and chest and one finds it difficult to breathe and cough. Additional symptoms are: dryness of breath, thick spittle, hoarseness of the voice, pain and heaviness of the head, restlessness, nausea and anxiety. (Note the difference: restlessness, nausea and anxiety occur more frequently with "measles" than with smallpox. On the other hand, pain in the back is more apparent with smallpox than with measles). Altogether one experiences heat over the whole body, one has an inflamed colon and one shows an overall shining redness, with a very pronounced redness of the gums. (Rhazes in Encyclopedia of Medicine).

- Meningitis:

Al-Razi compared the outcome of patients with meningitis treated with blood-letting with the outcome of those treated without it to see if blood-letting could help.

- Pharmacy:

Al-Razi contributed in many ways to the early practice of pharmacy by compiling texts, in which he introduces the use of "mercurial ointments" and his development of apparatus such as mortars, flasks, spatulas and phials. These were used in pharmacies until the early twentieth century.

- Ethics of medicine:

On a professional level, al-Razi introduced many practical, progressive, medical and psychological ideas. He attacked charlatans and fake doctors who roamed the cities and countryside selling their nostrums and "cures". At the same time, he warned that even highly educated doctors did not have the answers to all medical problems and could not cure all sicknesses or heal every disease, which was humanly speaking impossible. To become more useful in their services and truer to their calling, al-Razi advised practitioners to keep up with advanced knowledge by continually studying medical books and exposing themselves to new information.

He also wrote the following on medical ethics:

The doctor's aim is to do good, even to our enemies, so much more to our friends, and my profession forbids us to do harm to our kindred, as it is instituted for the benefit and welfare of the human race, and God imposed on physicians the oath not to compose mortiferous remedies.

- Incurable Diseases:

He made a distinction between curable and incurable diseases. Pertaining to the latter, he commented that in the case of advanced cases of cancer and leprosy the physician should not be blamed when he could not cure them. Al-Razi felt for physicians who took care for the wellbeing of princes, nobility, and women, because they did not obey the doctor's orders to restrict their diet or get medical treatment, thus making it most difficult being their physician.

Al-Razi's Books and articles on medicine:

- Al-Kitab al Hawi

This 23-volume set medical textbooks contains the foundation of gynecology, obstetrics and ophthalmic surgery.

- The Virtuous Life (al-Hawi الحاوي).

This monumental medical encyclopedia in nine volumes, known in Europe as The Large Comprehensive Liber (جامع الكبير), contains considerations and criticism on the Greek philosophers Aristotle and Plato, and expresses innovative views on many subjects. Because of this book alone, many scholars consider al-Razi the greatest medical scientist.

The al-Hawi is not a formal medical encyclopedia, but a posthumous compilation of al-Razi's working notebooks, which included original observations on diseases and therapies, based on his own clinical experience. It is significant since it contains a celebrated monograph on smallpox, the earliest one. It was translated into Latin in 1279 by Faraj ben Salim, a Sicilian physician employed by Charles of Anjou.

The al-Hawi also criticized the views of Galen, after al-Razi had observed many clinical cases which did not agree with Galen's descriptions of fevers. For example, he stated that Galen's descriptions of urinary ailments were in error; Galen had only seen three cases, while al-Razi had studied hundreds of such cases in hospitals of Baghdad and Rey.

- من لا يحضره الطبيب (Man la Yahduruhu Al-Tabib), "For One Who Has No Physician to Attend Him".

This is a medical adviser for the general public

Al-Razi was possibly the first doctor to deliberately write a home medical manual (remedial) directed at the general public. He dedicated it to the poor, the traveler, and the ordinary citizen who could consult it for treatment of common ailments when a doctor was not available. This book is of special interest to the history of pharmacy. Al-Razi described in its 36 chapters, diets and drug components that can be found in either an apothecary, a market place, in well-equipped kitchens, and in military camps. Thus, every intelligent person could follow its instructions and prepare the proper recipes with good results.

Some of the illnesses treated were headaches, colds, coughing, melancholy and diseases of the eye, ear, and stomach. For example, he prescribed for a feverish headache: "2 parts of duhn (oily extract) of rose, to be mixed with 1 part of vinegar; a piece of linen cloth is dipped in it and compressed on the forehead". He recommended as a laxative, "7 drams of dried violet flowers with 20 pears, macerated and well mixed, then strained. Add to this filtrate, 20 drams of sugar for a drink. In cases of melancholy, he invariably recommended prescriptions, which included either poppies or its juice (opium), Cuscuta epithymum (clover dodder) or both. For an eye-remedy, he advised myrrh, saffron, and frankincense, 2 drams each, to be mixed with 1 dram of yellow arsenic formed into tablets. Each tablet was to be dissolved in a sufficient quantity of coriander water and used as eye drops.

- Book for al-Mansur (Kitāb al-Manṣūrī)

Al-Razi dedicated this work to his patron Abū Ṣāliḥ al-Manṣūr, the Samanid governor of Ray. A Latin translation of it was edited in the 16th century by the Dutch writer Andreas Vesalius.

- Doubts about Galen (al-Shukūk ʿalā Jalīnūs)

Al-Razi rejects several claims made by the Greek physician, Galen, with respect to many of his cosmological and medical views. Al-Razi links medicine with philosophy, and states that sound practice demands independent thinking. He reports that Galen's descriptions do not agree with his own clinical observations regarding the run of a fever. He finds that his clinical experimental observations must override Galen's descriptions.

He criticized Galen's theory that the body possessed four separate liquid substances, whose balance are the key to health and a natural body-temperature. A sure way to upset such a system was to insert a liquid with a different temperature into the body resulting in an increase or decrease of bodily heat. Al-Razi noted that a warm drink would heat up the body to a degree much higher than its own natural temperature. Thus, the drink would trigger a response from the body, rather than transferring only its own warmth or coldness to it.

This line of criticism essentially had the potential to completely refute Galen's theory of four liquids, as well as Aristotle's theory of the four elements, on which it was grounded. Al-Razi's own alchemical experiments suggested other qualities of matter, such as "oiliness" and "sulphurousness", or inflammability and salinity, which were not readily explained by the traditional fire, water, earth, and air division of elements.

- The Diseases of Children

It was the first monograph to deal with pediatrics as an independent field of medicine.

Biographical Summary

Al-Razi was born in the city of Ray, also the origin of his name "al-Razi", and was a native speaker of Persian language. Ray was situated on the Great Silk Road that for centuries facilitated trade and cultural exchanges between East and West. It is located on the southern slopes of the Alborz mountain range situated near Tehran, Iran.

In his youth, al-Razi moved to Baghdad where he studied and practiced at the local bimaristan (hospital). Later, he was invited back to Rey by Mansur ibn Ishaq, then the governor of Ray, and became a bimaristan's head. He dedicated two books on medicine to Mansur ibn Ishaq: "The Spiritual Physic" and "Al-Mansūrī". Because of his newly acquired popularity as physician, al-Razi was invited to Baghdad where he assumed the responsibilities of a director in a new hospital named after its founder al-Muʿtaḍid (d. 902 CE). Under the reign of Al-Mutadid's son, Al-Muktafi (r. 902-908) al-Razi was commissioned to build a new hospital, which should be the largest of the Abbasid Caliphate. To pick the future hospital's location, al-Razi adopted what is nowadays known as an evidence-based approach, suggesting having fresh meat hung in various places throughout the city and to build the hospital where meat took longest to rot.

He spent the last years of his life in his native Rey suffering from glaucoma. His eye affliction started with cataracts and ended in total blindness. The cause of his blindness is uncertain. One account mentioned by Ibn Juljul attributed the cause to a blow to his head by his patron, Mansur ibn Ishaq, for failing to provide proof for his alchemy theories; while Abulfaraj and Casiri claimed that the cause was a diet of beans only. Allegedly, he was approached by a physician offering an ointment to cure his blindness. Al-Razi then asked him how many layers does the eye contain and when he was unable to receive an answer, he declined the treatment stating "my eyes will not be treated by one who does not know the basics of its anatomy".

The lectures of al-Razi attracted many students. As Ibn al-Nadim relates in his al-Fihrist, al-Razi was considered a shaikh, an honorary title given to one entitled to teach and surrounded by several circles of students. When someone raised a question, it was passed on to students of the 'first circle'; if they did not know the answer, it was passed on to those of the 'second circle', and so on. When all students would fail to answer, al-Razi himself would consider the query. Al-Razi was a generous person by nature, with a considerate attitude towards his patients. He was charitable to the poor, treated them without payment in any form, and wrote for them a treatise "Man La Yaḥḍuruhu al-Ṭabīb", or Who Has No Physician to Attend Him, with medical advice. One former pupil from Tabaristan came to look after him, but as al-Biruni wrote, al-Razi rewarded him for his intentions and sent him back home, proclaiming that his final days were approaching. According to Biruni, al-Razi died in Rey in 925 sixty years of age. Biruni, who considered al-Razi his mentor, was among the first who penned a short biography of al-Razi including a bibliography of his numerous works.

Ibn al-Nadim recorded an account by al-Razi of a Chinese student who copied down all of Galen's works in Chinese as al-Razi read them to him out loud after the student learned fluent Arabic in 5 months and attended al-Razi's lectures.

After his death, his fame spread beyond the Middle East to Europe, and lived on. In an undated catalog of the library at Peterborough Abbey, most likely from the 14th century, al-Razi is listed as a part author of ten books on medicine.

Al-Razi died in Rey in 925 AD, at sixty years of age.

Remarks

The scientific approach and preference to experimental facts, led al-Razi to express his thoughts in religion, that did not sit well with his contemporaries. He was accused, even by such learned people as Al-Biruni and Ibn Rushd, for his thinking about the nature of Prophethood and Revealed knowledge. In response, al-Razi stated the following in his Philosophical Approach:

"(...) In short, while I am writing the present book, I have written so far around 200 books and articles on different aspects of science, philosophy, theology, and hekmat (wisdom). (...) I never entered the service of any king as a military man or a man of office, and if I ever did have a conversation with a king, it never went beyond my medical responsibility and advice. (...) Those who have seen me know, that I did not into excess with eating, drinking or acting the wrong way. As to my interest in science, people know perfectly well and must have witnessed how I have devoted all my life to science since my youth. My patience and diligence in the pursuit of science has been such that on one special issue specifically I have written 20,000 pages (in small print), moreover I spent fifteen years of my life -night and day- writing the big collection entitled Al Hawi. It was during this time that I lost my eyesight, my hand became paralyzed, with the result that I am now deprived of reading and writing. Nonetheless, I've never given up, but kept on reading and writing with the help of others. I could make concessions with my opponents and admit some shortcomings, but I am most curious, what they have to say about my scientific achievement. If they consider my approach incorrect, they could present their views and state their points clearly, so that I may study them, and if I determined their views to be right, I would admit it. However, if I disagreed, I would discuss the matter to prove my standpoint. If this is not the case, and they merely disagree with my approach and way of life, I would appreciate they only use my written knowledge and stop interfering with my behavior."

— Al-Razi, The Philosophical Approach

Annotations

Apart from being a leading figure in medicine, and a prominent philosopher, al-Razi also was an accomplished chemist.

In philosophy, Al-Razi's metaphysical doctrine derives from the theory of the "five eternals", according to which the world is produced out of an interaction between God and four other eternal principles: soul, matter, time, and place. This level of sophistication of thought is unparalleled; and definitely way exceeds the relatively crude Aristotelian thought using fire, air, earth, and water.

Half a century after the death of al-Razi, Ibn an-Nadim attributed a series of twelve books to al-Razi, plus an additional seven, including his refutation to al-Kindi's denial of the validity of alchemy. Al-Kindi (801–873 CE) had been appointed by the Abbasid Caliph, Ma'mun, founder of Baghdad, to 'the House of Wisdom'. He was a philosopher and an opponent of alchemy. Al-Razi's two best-known alchemical texts are: al-Asrar (الاسرار) "The Secrets", and Sirr al-Asrar (سر الاسرار) "The Secret of Secrets", which incorporate the gist of his work on alchemy.

Apparently al-Razi's contemporaries believed that he had obtained the secret of turning iron and copper into gold. Biographer Khosro Moetazed reports in Mohammad Zakaria Razi that a certain General Simjur confronted al-Razi in public, and asked whether that was the underlying reason for his willingness to treat patients without a fee. "It appeared to those present that al-Razi was reluctant to answer; he looked sideways at the general and replied":

I understand alchemy and I have been working on the characteristic properties of metals for an extended time. However, it still has not turned out to be evident to me, how one can transmute gold from copper. Despite the research from the ancient scientists done over the past centuries, there has been no answer. I very much doubt if it is possible...

11. Ibn al-Jazzar

Ahmed Bin Jaafar Bin Brahim Ibn Al Jazzar Al-Qayrawani

(Arabic: أبو جعفر أحمد بن أبي خالد بن الجزار القيرواني),

(895–979),

was a physician and a writer on medical science.

Scientific Contributions

Al Jazzar wrote on medicine: "Zad Al Mussafir".

It is translated into Latin, Greek and Hebrew, it has been copied, recopied, and printed in France and Italy in the sixteenth century.

This book is influenced by Ibn Sina. It was adopted and popularized in Europe as a book for education in medicine.

It is a medicine handbook from head to toe, designed for clinical teaching. There are lessons written after the course, as noted by the author in the conclusion of his book. The author names the disease, lists the known symptoms, gives the treatment and provides the prognosis.

Unlike European scientists who took the work of Muslim scientists without acknowledgement, Ibn Al Jazzar observed intellectual integrity and cited in references the names of foreign authors.

Ibn Al Jazzar had additional works:

• books on geriatric medicine

- health of elderly (Kitāb Ṭibb al-Mashāyikh) or (Ṭibb al-Mashāyikh wa-ḥifẓ ṣiḥḥatihim).

- book on sleep disorders and another one on forgetfulness and how to strengthen memory (Kitāb al-Nisyān wa-Ṭuruq Taqwiyat al-Dhākira)

- Treatise on causes of mortality (Risāla fī Asbāb al-Wafāh).

- Multiple books on pediatrics, fevers, sexual disorders, medicine of the poor, therapeutics, coryza, stomach disorders, leprosy, separate drugs, compound drugs.

In addition to his works in medical sciences, he has works in other areas of science, e.g., history, animals and literature. Ibn Al Jazzar wrote a number of books. Some deal with grammar, history, jurisprudence, prosody, etc. Many of these books, quoted by different authors are lost.

Biographical Summary

Ibn Juljul, an Andalusian physician had a student named Ibn Bariq. He went to Qayrawan, Tunisia, to learn medicine and gives the following account.

Ahmed Ben Jaafar Ben Brahim Ibn Al Jazzar was born in Qayrawan around 895, and died around 979 AD. He lived for about 84 years. He was married, although he did not have children. He had learned the Quran at Kuttab in his youth; grammar, theology, fiqh and history at the mosque Okba Ibn Nafaa.

He had learned medicine from his father and his uncle who were physicians, and from Ishaq Ibn Suleiman who was also a physician in Qayrawan.

Ibn Al Jazzar had an extensive library.

Ibn Al Jazzar was calm and quiet. He did not attend funerals or weddings, and did not take part in festivities. He had great respect for himself. He avoided compromises, did not attend the court and members of the regime, thus taking on Fouqaha example of the time. This may explain the fact that when he treated the son of Cadhi Al Nooman, he refused to receive as a gift a costume of 300 mithkals.

It is also by respect for the Emir that he had not realized his desire to visit Andalusia, the relationship between the two governments of Mahdia and Cordoba were tense. It is also by respect for the Emir that he did not begin his pilgrimage to Mecca in spite of his strong desire to do so. The Emir was Shia and for ceremonial purposes and policies, he creates barriers to pilgrims and forced them to pass through Mahdia and pay a toll.

But he went every Friday to Mahdia to the uncle of the Emir El Moez Lidin Allah, which he was bound by friendship. During the heat of the summer, he went to Monastir and lived in ribat with valiant soldiers who watched the boundaries. Ibn Al Jazzar preparing himself the medicines and had an assistant serve them who stood in the vestibule of the house, and who collected the fees of the consultations.

Ibn Al Jazzar left 24,000 gold dinars at his death. The Aghlabid dinar weighed 4.20 grams.

He was born in 895 AD in Qayrawan in modern-day Tunisia. He died in 979 AD.

12.   Al-Tamimi

Muhammad ibn Sa'id al-Tamimi

(Arabic: أبو عبد الله محمد بن سعيد التميمي),

known by his kunya, "Abu Abdullah," but more commonly as Al-Tamimi,

(died 990),

was a physician, renowned for his medical works.

Note: To disambiguate, another scientist Muḥammad ibn Umayl al-Tamīmī (Arabic: محمد بن أميل التميمي), was an Egyptian chemist who lived from c. 900 to c. 960 AD.

Scientific Contributions

Al-Tamimi wrote a medical work for the vizier, Ya'qub ibn Killis (930–991).

He was especially known for his having concocted a theriac, a proven antidote in snakebite and other poisons. Al-Tamimi named it "tiryaq al-fārūq". He specialized in compounding drugs and medicines.

Al-Tamimi's wrote a treatise: The Guidebook to Basics in Food Nutrition and the Properties of non-compounded Medicines (Arabic: كتاب المرشد الى جواهر الأغذيه وقوت المفردات من الأدويه), known also under its abbreviated name, Al-Murshid. This work which researches the properties of certain plants (antidotes) and minerals has laid the foundation for subsequent works written on medicine by other authors. Among them is the work of Ibn al-Baytar in Cairo (d. 1248/646 H), in which he treats on various antidotes used to remedy poisons inflicted by snakebite and scorpion stings. Another one is the work of `Ali ibn `Abd al-`Aẓim al-Anṣāri in Syria (1270), which is a treatise on antidotes for poisons entitled "Dhikr al-tiryaq al-faruq", where he quotes from al-Tamimi's works (some of which are no longer extant). Maimonides (1138–1204), also made use of his works.

Although only portions of al-Tamimi's seminal work have survived, a section of the book (chapters 12, 13 and 14) which discusses rocks and minerals, including asphalt, is today held at the Bibliothèque nationale de France in Paris, in manuscript form (Paris MS. no. 2870), consisting of 172 pages.

Other sections of al-Tamimi's original work were copied by `Ali ibn `Abd al-`Aẓim al-Anṣāri in 1270, now preserved at the U.S. National Library of Medicine in Bethesda, Maryland.

Ibn al-Baytar, cites al-Tamimi some seventy times. Abstracts of these manuscripts have been published in Hebrew by Yaron Serri and Zohar Amar of Bar-Ilan University, in the book, "The Land of Israel and Syria as Described by al-Tamimi."

Al-Tamimi's works on materia medica are an invaluable source for understanding the curative remedies that were in use in Syria and Palestine. They often relate to the daily life and beliefs, as well as the practical usages of plants. Al-Tamimi's theriac recension is of particular importance to botanists, as he describes in great detail the recognizable features of the plants used as electuaries, as also the season for gathering such plants.

His other important medical works include:

- Māddat-ul-Baqā' fi Iṣlāḥ Fasad il-Ḥawā w-al-taḥarruz min Ḍarar-il-Awbā` (The Extension of Life by Purifying the Air of Corruption and Guarding against the Evil Effects of Pestilences), a book written for his friend, the Fatimid vizier, Ya'qub ibn Killis.

- Maqālah fi Māhīyat-ul-Ramad wa Anwā'uhū wa Asbābuhū wa 'Ilājuh (Treatise on the Nature of Ophthalmology and its Types, Causes and Treatment).

- Ḥabīb al-'arūs, wa-rayḥān al-nufūs (The Beloved of the Bride, and the [Fragrant] Basil of Souls).

- Miftāḥ al-Surrūr fi kul al-Hummūm (The Key to Pleasure in all Worries).

- Several works on how to compound Theriac.

Biographic Summary

Born in Jerusalem, Al-Tamimi spent his early years in and around Jerusalem. Al-Tamimi possessed an uncommon knowledge of plants and their properties, such that his service in this field was highly coveted and brought him to serve as the personal physician of the Ikhshidid Governor of Ramla, al-Hassan bin Abdullah bin Tughj al-Mastouli. He serviced them in Old Cairo, Egypt.

Around 970, Al-Tamimi had settled in Old Cairo, Egypt, and prospered in his medical field.

He died in 990 AD.

13. Al-Jabali

Abu Abd Allah Mohammed ibn Abdun al-Jabali al-Adadi

(Arabic: محمد بن عبدون الجبلي العذري),

(died after 976),

was a physician and mathematician from Al-Andalus.

Scientific Contributions

Al-Jabali was in charge of the hospital in Cairo.

Biographical Summary

Al-Jabali was from Al-Andalus. He travelled to the learning centers in the East in the years after 958 C.E. He stayed in Basra and visited al-Fustat (Old Cairo) where he was put in charge of the hospital.

Ibn Jabali studied the ideas of Abu Sulayman Sijistani (d. 990) and according to one source he met him personally in Basra.

Ibn Jabali returned to Cordoba in 971 C.E.

He entered the service of the Caliph al-Mustansir and his son Hisham II al-Mu'ayad.

Ibn Jabali was the teacher of Ibn al-Kattani.

Ibn Jabali died sometime after 976 CE

14.   Ammar al-Mawsili

Abu al-Qasim Ammar ibn Ali al-Mawsili

(Arabic: عمار الموصلي),

was an eleventh-century ophthalmologist. Despite little being known about his life, he is acknowledged as the most original of ophthalmologists.

Scientific Contributions

Ammar al-Mawsili is the inventor of a hypodermic syringe, which he used to remove cataracts, a major cause of blindness.

Regarding his invention he wrote the following:

Then I constructed the hollow needle, but I did not operate with it on anybody at all, before I came to Tiberias. There came a man for an operation who told me: Do as you like with me, only I cannot lie on my back. Then I operated on him with the hollow needle and extracted the cataract; and he saw immediately and did not need to lie, but slept as he liked. Only I bandaged his eye for seven days. With this needle nobody preceded me. I have done many operations with it in Egypt.

Biographical Summary

As his nisba indicates, Ammar was born in Mosul, and later moved to Egypt, where he settled during the reign of the Fatimid caliph al-

Hakim bi-Amr Allah, to whom he wrote his composition: Kitāb al-muntakhab fi ilm al-ayn ("The book of choice in ophthalmology").

He was a contemporary of the oculist Ali ibn Isa.

15.   Ibn Juljul

Abu Dawud Sulayman ibn Hassan Ibn Juljul

(Arabic: سليمان بن حسان ابن جلجل),

(c. 944 Córdoba – c. 994),

was an Andalusian physician and pharmacologist.

Scientific Contributions

He wrote treatise on the history of medicine. His works on pharmacology were frequently quoted by physicians in Al-Andalus.

His works were also studied by Albertus Magnus, but were wrongly attributed to a Gilgil, as the Europeans are generally cavalierly careless about Arab names, and their scientific research contributions.

Ibn Juljul's major book is Ṭabaqāt al-aṭibbā' w'al-hukamā' (Generations of physicians and Wise Men, Arabic: طبقات الأطباء والحكماء) which is an important work on the history of medicine. The book includes 57 biographies of famous physicians. The included biographies of contemporary Andalucian physicians auxiliarily provide insight about life in Cordoba. One of the biographies is that of Mohammed ibn Abdun al-Jabali, his colleague physician at the court of Cordoba. Composed in 987, the Ṭabaqāt is considered to be the second oldest collection of biographies of physicians; the earliest being Ta'rīkh al-aṭibbā' by Ishaq ibn Hunayn.

Ibn Juljul states that: scholars appear only in states whose kings seek knowledge.

Ibn Juljul wrote a number of additional treatises on pharmacology.

Biographical Summary

Ibn Juljul was born in 944 AD in Andalusia, and he died in 994.

Starting from the age of fourteen, Ibn Juljul studied medicine for ten years working under the physician Hasdai ibn Shaprut.

He later became the personal physician of Caliph Hisham II, and continued working as a teacher of medicine.

Ibn al-Baghunish of Toledo was one of his disciples.

16. Ali ibn Isa al-Kahhal

ʿAlī ibn ʿĪsā al-Kahhal

(Arabic: علي بن عيسى الكحال), surnamed "the oculist" (al-kahhal),

(fl. 1010 AD),

was a celebrated ophthalmologist.

Scientific Contributions

ʿAlī ibn ʿĪsā al-Kahhal was the author of the influential Memorandum of the Ophthalmologists, where for the first time a surgical anesthetic procedure is prescribed. The book encompassed information on treatment and classification of over one hundred different eye diseases. In the book, eye diseases were sorted by their anatomical location. The Notebook of the Oculists was widely used by European physicians for hundreds of years.

ʿAlī ibn ʿĪsā al-Kahhal was the first to discover the symptoms of Vogt–Koyanagi–Harada syndrome (VKH) - ocular inflammation associated with a distinct whitening of the hair, eyebrows, and eyelashes. He was also the first to classify epiphora as being a result of overzealous cautery of pterygium.

In addition to this pioneering research, ʿAlī ibn ʿĪsā al-Kahhal prescribed treatments for epiphora based on the stage of the disease – namely treatment in the early stages with astringent materials, for example ammonia salt, burned copper, or lid past and a hook dissection with a feathered quill for chronic stages of epiphora.

ʿAlī ibn ʿĪsā al-Kahhal is also the first to describe temporal arteritis, although Sir Jonathan Hutchinson (1828–1913) is erroneously credited with this.

It is normal practice among the European science community to deliberately not acknowledge the scientific research of the Muslim Scientists, generally dismissing it as an exercise in the translation of the Greek sciences. They freely and cavalierly violate the academic norms and the scientific integrity. They always ascribe the scientific discoveries to their own members, so much so that today their text books carry no references to the pioneering research of the Muslim Scientists.

Biographical Summary

ʿAlī ibn ʿĪsā al-Kahhal, surnamed "the Ophthalmologist" (al-kahhal), was the best known and most celebrated ophthalmologist. He is known in Europe as Jesu Occulist.

17.   Al-Zahrawi

Abū al-Qāsim Khalaf ibn al-'Abbās al-Zahrāwī al-Ansari

(Arabic: أبو القاسم خلف بن العباس الزهراوي),

popularly known as Al-Zahrawi (الزهراوي),

(936–1013),

was a physician, surgeon and chemist from Andalusia. He is the "father of modern surgery".

Scientific Contributions

Al-Zahrawi's principal work is the Kitab al-Tasrif, a thirty-volume encyclopedia of medical practices. The surgery chapter of this work was later translated into Latin, attaining popularity and becoming the standard textbook in Europe for the next five hundred years.

Al-Zahrawi's pioneering contributions to the field of surgical procedures and instruments had an enormous impact in the East and West well into the modern period, where some of his discoveries are still applied in medicine to this day. He pioneered the use of catgut for internal stitches, and his surgical instruments are still used today.

He was the first physician to identify the hereditary nature of hemophilia and describes an abdominal pregnancy, a subtype of ectopic pregnancy.

He was the first to discover the root cause of paralysis.

He developed surgical devices for Caesarean sections and cataract surgeries.

Al-Zahrawi specialized in curing disease by cauterization. He invented several devices used during surgery, for purposes such as inspection of the interior of the urethra and also inspection, applying and removing foreign bodies from the throat, the ear and other body organs. He was also the first to illustrate the various cannulae and the first to treat a wart with an iron tube and caustic metal as a boring instrument.

In tracheotomy, he did treat a slave girl who had cut her own throat in a suicide attempt. Al-Zahrawi sewed up the wound and the girl recovered, thereby proving that an incision in the larynx could heal. In describing this important case-history he wrote:

A slave-girl seized a knife and buried it in her throat and cut part of the trachea; and I was called to attend to her. I found her bellowing like a sacrificial animal that has had its throat cut. So I laid the wound bare and found that only a little haemorrhage had come from it; and I assured myself that neither an artery nor jugular vein had been cut, but air passed out through the wound. So, I hurriedly sutured the wound and treated it until healed. No harm was done to the slave-girl except for a hoarseness in the voice, which was not extreme, and after some days she was restored to the best of health. Hence, we may say that laryngotomy is not dangerous.

Al-Zahrawi also pioneered neurosurgery and neurological diagnosis. He is known to have performed surgical treatments of head injuries, skull fractures, spinal injuries, hydrocephalus, subdural effusions and headache. The first clinical description of an operative procedure for hydrocephalus was given by Al-Zahrawi who clearly describes the evacuation of superficial intracranial fluid in hydrocephalic children.

Al-Zahrawi's thirty-volume medical encyclopedia, Kitāb al-Taṣrīf, completed in the year 1000, covered a broad range of medical topics, including on surgery, medicine, orthopedics, ophthalmology, pharmacology, nutrition, dentistry, childbirth, and pathology.

The first volume in the encyclopedia is concerned with general principles of medicine, the second with pathology, while much of the rest discuss topics regarding pharmacology and drugs. The last treatise and the most celebrated one is about surgery. Al-Zahrawi stated that he chose to discuss surgery in the last volume because surgery is the highest form of medicine, and one must not practice it until he becomes well-acquainted with all other branches of medicine.

The work contained data that had accumulated during a career that spanned almost 50 years of training, teaching and practice. In it he also wrote of the importance of a positive doctor-patient relationship and wrote affectionately of his students, whom he referred to as "my children". He also emphasized the importance of treating patients irrespective of their social status. He encouraged the close observation of individual cases in order to make the most accurate diagnosis and the best possible treatment.

Not always properly credited, modern evaluation of Kitab al-Tasrif manuscript has revealed on early descriptions of some medical procedures that were ascribed to later physicians. For example, Al-Zahrawi's Kitab al-Tasrif described both what would later become known as "Kocher's method" for treating a dislocated shoulder and "Walcher position" in obstetrics. Moreover, the Kitab al-Tasrif described how to ligature blood vessels almost 600 years before Ambroise Paré, and was the first recorded book to explain the hereditary nature of haemophilia. It was also the first to describe a surgical procedure for ligating the temporal artery for migraine, also almost 600 years before Pare recorded that he had ligated his own temporal artery for headache that conforms to current descriptions of migraine. Al-Zahrawi was, therefore, the first to describe the migraine surgery procedure that is enjoying a revival in the 21st century, spearheaded by Elliot Shevel a South African surgeon.

This is a common European practice: to do the utmost to not acknowledge the scientific achievements of Muslim Scientists; and instead to attribute the research achievements of the Muslim scientists to their own European agents who came not weeks, years, decades but centuries later. There is no nice way to describe this type of academic dishonesty by the Europeans and an utter lack of their research integrity. This is the systemic behavior of European academicians and researchers, even if there exist some exceptions here and there, such as the quotes by Donald Campbell and Guy de Chauliac given at the end of this subsection.

On Surgery and Instruments is the 30th and last volume of the Kitab al-Tasrif. It was the one which established his authority in Europe for centuries to come.

It was the first illustrated surgical guide ever written. Its contents and descriptions have contributed in many technological innovations in medicine, notably which tools to use in specific surgeries. In his book, al-Zahrawi draws diagrams of each tool used in different procedures to clarify how to carry out the steps of each surgery. The full text consists of three books, intended for medical students looking forward to gaining more knowledge within the field of surgery regarding procedures and the necessary tools.

The book was translated into Latin in the 12th century by Gerard of Cremona. It soon found popularity in Europe and became a standard text in all major Medical universities like those of Salerno and Montpellier. It remained the primary source on surgery in Europe for the next 500 years.

Noting the importance of anatomy al-Zahrawi wrote:

"Before practicing surgery, one should gain knowledge of anatomy and the function of organs so that he will understand their shape, connections and borders. He should become thoroughly familiar with nerves muscles bones arteries and veins. If one does not

59

comprehend the anatomy and physiology one can commit a mistake which will result in the death of the patient. I have seen someone incise into a swelling in the neck thinking it was an abscess, when it was an aneurysm and the patient dying on the spot."

In urology, al-Zahrawi wrote about taking stones out of the bladder. By inventing a new instrument, an early form of the lithotrite which he called "Michaab", he was able to crush the stone inside the bladder without the need for a surgical incision. His technique was important for the development of lithotomy, and an improvement over the existing techniques in Europe which caused severe pain for the patient, and came with high death rates.

In dentistry and periodontics, al-Zahrawi had the most significant contribution out of all physicians, and his book contained the earliest illustrations of dental instruments. He was known to use gold and silver wires to ligate loosened teeth, and has been credited as the first to use replantation in the history of dentistry. He also invented instruments to scale the calculus from the teeth, a procedure he recommended as a prevention from periodontal disease.

Al-Zahrawi introduced over 200 surgical instruments, which include, among others, different kinds of scalpels, retractors, curettes, pincers, specula, and also instruments designed for his favored techniques of cauterization and ligature.

He also invented hooks with a double tip for use in surgery. Many of these instruments were newly invented, never previously used.

His use of catgut for internal stitching is still practiced in modern surgery. The catgut appears to be the only natural substance capable of dissolving and is acceptable by the body. An observation Al-Zahrawi discovered after his monkey ate the strings of his oud. This mindset is that of a true natural scientist.

Al-Zahrawi also invented the forceps for extracting a dead fetus, as illustrated in the Kitab al-Tasrif.

"On cauterization for numbness", al-Zahrawi declares the procedure "should not be attempted except by one who has a good knowledge of the anatomy of the limbs and of the exits of the nerves that move the body". He warns that another procedure should not be attempted by any surgeon lacking "long training and practice in the use of cautery". He is not afraid to depart from old practice, disparaging the opinions that cauterization should only be used in the spring or that gold is the best material for cauterization: "cauterization is swifter and more successful with iron". In "On cauterization for pleurisy", he notes that the introduction of a red-hot probe into the intercostal space to evacuate pus from an abscess could result in the creation of "an incurable fistula" or even the immediate death of the patient.

In pharmacy and pharmacology, Al-Zahrawi pioneered the preparation of medicines by sublimation and distillation. He dedicated the 28th chapter of his book to pharmacy and pharmaceutical techniques. The chapter was later translated into Latin under the title of Liber Servitoris, where it served as an invaluable source for Europeans. The book is of particular interest, as it provides the reader with recipes and explains how to prepare the "simples" from which were compounded the complex drugs.

Al-Zahrawi also touched upon the subject of cosmetics and dedicated a chapter for it in his medical encyclopedia. The treatise was translated into Latin. Al-Zahrawi considered cosmetics a branch of medicine, which he called "Medicine of Beauty" (Adwiyat al-Zinah). He deals with perfumes, scented aromatics and incense. He also invented perfumed sticks rolled and pressed in special molds, perhaps the earliest antecedents of present-day lipsticks and solid deodorants.

Donald Campbell, a historian of Arabic medicine, described Al-Zahrawi's influence on Europe as follows:

al-Zahrawi's lucidity and method of presentation awakened a prepossession in favor of Arabic literature among the scholars of the West: the methods of Albucasis eclipsed those of Galen and maintained a dominant position in medical Europe for five hundred years. He, however, helped to raise the status of surgery in Christian Europe; in his book on fractures and luxations, he states that 'this part of surgery has passed into the hands of vulgar and uncultivated minds, for which reason it has fallen into contempt.' The surgery of Albucasis became firmly grafted on Europe after the time of Guy de Chauliac (d.1368).

In the 14th century, the French surgeon Guy de Chauliac quoted al-Tasrif over 200 times. Pietro Argallata (d. 1453) described Al-Zahrawi as "without doubt the chief of all surgeons". Al-Zahrawi's influence continued for at least five centuries, extending well into the Renaissance, as evidenced by al-Tasrif's frequent reference by French surgeon Jacques Daléchamps (1513–1588).

The street in Cordova where he lived is named in his honor as "Calle Albucasis". On this street he lived in house no. 6, which is preserved today by the Spanish Tourist Board with a bronze plaque (since January 1977) which reads: "This was the house where Al-Zahrawi lived."

Biographical Summary

Al-Zahrawi was born in the city of Azahara, 8 kilometers northwest of Cordoba, Andalusia. His birth date is not known for sure, however, scholars agree that it was after 936, the year his birthplace city of Azahara was founded. The nisba (attributive title), Al-Ansari, in his name, suggests origin from the Medina.

Al-Zahrawi died in 1013 AD.

He lived most of his life in Cordoba. It is also where he studied, taught and practiced medicine and surgery until shortly before his death in about 1013, two years after the sacking of Azahara.

Few details remain regarding his life, aside from his published work, due to the destruction of El-Zahra during Castillian attacks. His name first appears in the writings of Abu Muhammad bin Hazm (993–1064), who listed him among the greatest physicians. There is a more detailed biography of al-Zahrawī in al-Ḥumaydī's Jadhwat al-Muqtabis, completed six decades after al-Zahrawi's death.

Al-Zahrawi was a court physician to the Andalusian caliph Al-Hakam II. He was a contemporary of Andalusian chemists such as Ibn al-Wafid, al-Majriti and Artephius.

He devoted his entire life and genius to the advancement of medicine as a whole and surgery in particular.

18. Ibn al-Kattani

Abu Abd Allah Muhammad ibn al-Husayn Ibn al-Kattani al-Madhiji

(Arabic: ابن الكتاني),

sometimes nicknamed "al-Mutatabbib" (the physician),

(951–1029),

was a well-known polymath scholar, physician, philosopher, astrologer, man of letters, and poet.

Scientific Contributions

Ibn al-Kattani wrote books on logic, inference and deduction.

For some time he was the personal physician of Al-Mansur Ibn Abi Aamir, sultan of al-Andalus, and wrote the treatise titled "The Treatment of Dangerous Diseases Appearing Superficially on the Body" (Mu`alajat al-amrad al-khatirah al-badiyah `ala al-badan min kharij). It was cited by later writers, but thought to be now lost, until a copy of it was discovered among the manuscripts now at the National Library of Medicine.

Much of the treatise is on the subject of poisonous bites.

Al-Kattani also wrote an anthology of Andalusian poetry.

He is also famous for his book on metaphor in Andalusian poetry.

Biographical Summary

Ibn al-Kattani was born in 951 AD in Córdoba in the Caliphate of Cordoba. He died in Saragossa in 1029.

19. Ibn al-Thahabi

Abu Mohammed Abdellah Ibn Mohammed Al-Azdi

(Arabic: ابو محمد عبدالله بن محمد الأزدي),

known also as Ibn Al-Thahabi or Ibn al-Zahabi,

(ca. ? - 1033 CE),

was a physician, famous for writing the first known alphabetical encyclopedia of medicine.

Scientific Contributions

Ibn Al-Thahabi wrote a comprehensive treatise on "Kitab Al-Ma'a" (The Book of Water), a medical encyclopedia that lists the names of diseases, medicines, physiological processes, and treatments.

In this encyclopedia, Ibn Al-Thahabi not only lists the names but adds numerous original ideas about the function of the human organs. The book also contains an array of herbal treatments and a course for the treatment psychological symptoms.

Ibn Al-Thahabi's medical doctrine is that cure must start from controlled food and exercise; if it persists then use specific individual medicines; if it still persists then use medical compounds; and if the disease continues, surgery is performed.

Biographical Summary

Ibn Al-Thahabi was born in Suhar, Oman. He moved to Basra, then to Persia where he studied under Al-Biruni and Ibn Sina. Later he migrated to Jerusalem and finally settled in Valencia, in Al-Andalus.

Ibn Al-Thahabi died in 1033 AD.

20. Ibn Sina

Abū ʿAlī al-Ḥusayn bin ʿAbdullāh ibn al-Ḥasan bin ʿAlī bin Sīnā al-Balkhi al-Bukhari

(mistakenly known in the West as Avicenna due to corruption of his name in the West),

(Arabic: أبو علي الحسين بن عبد الله بن الحسن بن علي بن سينا البلخي البخاري),

(980 – June 1037 CE),

is the father of modern medicine. He was a polymath, a most significant physician, astronomer, philosopher, and writer.

Scientific Contributions

Of the 450 works he is believed to have written, around 240 have survived, including 150 on philosophy and 40 on medicine.

His most famous works are "The Book of Healing", which is a philosophical and scientific encyclopedia, and "The Canon of Medicine", which is a medical encyclopedia which became a standard medical text at many universities and remained in use till 1650. Besides philosophy and medicine, Ibn Sina wrote treatises on astronomy, alchemy, geography, geology, psychology, Islamic theology, logic, mathematics, physics, and works of poetry.

His Book of Healing earned him the title of being the father of medical sciences. It was also part of the Muslim scientists' work that stirred Renaissance in Europe. The Book of Healing became available in Europe in partial Latin translation some fifty years after its composition, under the title Sufficientia. Some authors have identified a "Latin Avicennism" as flourishing for some time, paralleling the more influential Latin Averroism, but suppressed by the Parisian (French) decrees of 1210 and 1215. William of Auvergne, Bishop of Paris and Albertus Magnus, was greatly influenced by the work of Ibn Sina in psychology, and theory of knowledge. The thoughts of Thomas Aquinas were formed by the works of Ibn Sina in metaphysics. This was the time when the Muslim enlightenment was beginning to filter through the Dark Ages of Europe.

Ibn Sina's Principal works

Al-Qanun fi't-Tibb

Ibn Sina authored a five-volume medical encyclopedia: The Canon of Medicine (Al-Qanun fi't-Tibb). It was used as the standard medical textbook in Europe up to the 18th century. The Canon still plays an important role in Unani medicine, the herbal approach to cure.

Ibn Sina considered whether events like rare diseases or disorders have natural causes. He used the example of polydactyly to explain that causal reasons exist for all medical events. This view of medical phenomena, and sciences in general, is the foundational basis for the future developments, including the Renaissance in Europe.

Kitab al-shifa

The Book of Healing is a not just about medicine, rather it is all-encompassing approach to human health; including also philosophy of science, logic, psychology, physics, and geology. Following are major topics that it encompasses in great details.

- Earth sciences

Ibn Sina wrote on Earth sciences such as geology in The Book of Healing. While discussing the formation of mountains, he explained:

Either they are the effects of upheavals of the crust of the earth, such as might occur during a violent earthquake, or they are the effect of water, which, cutting itself a new route, has denuded the valleys, the strata being of different kinds, some soft, some hard ... It would require a long period of time for all such changes to be accomplished, during which the mountains themselves might be somewhat diminished in size.

- Philosophy of science

In the Al-Burhan (On Demonstration) section of The Book of Healing, Ibn Sina discussed the philosophy of science and described a scientific method of inquiry. He discussed Aristotle's Posterior Analytics and significantly diverged from it on several points. Ibn Sina discussed the issue of a proper methodology for scientific inquiry and the question of "How does one acquire the first principles of a science?" He asked how a scientist would arrive at "the initial axioms or hypotheses of a deductive science without inferring them from some more basic premises?" He explained that the ideal situation is when one grasps that a "relation holds between the terms, which would allow for absolute, universal certainty". Ibn Sina then added two further methods for arriving at the first principles: the ancient Aristotelian method of induction (istiqra), and the method of examination and experimentation (tajriba). Ibn Sina criticized Aristotelian approach using "induction", arguing that "it does not lead to the absolute, universal, and certain premises that it purports to provide." In its place, he developed a "method of experimentation as a means for scientific inquiry."

- Logic

A formal system of temporal logic was studied by Ibn Sina. Ibn Sina discoursed on the study of the relationship between temporalis and the implication. Ibn Sina's work was further developed by Najm al-Dīn al-Qazwīnī al-Kātibī and became the dominant system of logic. Ibn Sina's logic was used by European logicians such as Albertus Magnus and William of Ockham. Ibn Sina endorsed the law of non-contradiction, that a fact could not be both true and false at the same time and in the same sense of the terminology used. He suggested for the truth of the law of non-contradiction that "Anyone who denies the law of non-contradiction if beaten and burned would realize that to be beaten is not the same as not to be beaten, and to be burned is not the same as not to be burned."

- Physics

In mechanics, Ibn Sina, in The Book of Healing, developed a theory of motion, in which he made a distinction between the inclination (tendency to motion) and force of a projectile, and concluded that motion was a result of an inclination (mayl) transferred to the projectile by the thrower, and that projectile motion in a vacuum would not cease. He viewed inclination as a permanent force whose effect is dissipated by external forces such as air resistance.

Please note that this enunciation is an enunciation of the laws of motion, erroneously attributed to Newton; or the theory of impetus attributed to Buridan in the 14th century.

It is a by now a familiar deception technique by the European writers to require absolute certainty in such historical happenings by injecting doubt via making statements like "it is unclear if Buridan was influenced by Ibn Sina, directly." In other instances the same European writers are happy with purely conjectural thoughts of their own that suggest, without even pretending to seek for evidence, that the Muslim scientists simply just translated what the Greeks had accomplished; thereby the European academicians and researchers omit to cite any Muslim references in their works, and they jump directly from Greek references to European references. This is systemic, dishonest, and gravely erroneous lacking research integrity.

In optics, Ibn Sina was among those who argued that light had a speed, observing that "if the perception of light is due to the emission of some sort of particles by a luminous source, the speed of light must be finite." He also provided an explanation of the rainbow phenomenon.

In 1253, a Latin text entitled Speculum Tripartitum stated the following regarding Ibn Sina's theory on heat: Ibn Sina says in his book of heaven and earth, that heat is generated from motion in things.

- Psychology

Ibn Sina's legacy in classical psychology is primarily embodied in the Kitab al-nafs parts of his Kitab al-shifa (The Book of Healing) and Kitab al-najat (The Book of Deliverance). These books were known in Latin under the title De Anima.

Ibn Sina's psychology requires that connection between the body and soul be strong enough to ensure the soul's individuation, but weak enough to allow for its immortality. Ibn Sina grounds his psychology on physiology, which means his account of the soul is one that deals almost entirely with the natural science of the body and its abilities of perception. Thus, the philosopher's connection between the soul and body is explained almost entirely by his understanding of perception; in this way, bodily perception interrelates with the immaterial human intellect. In sense perception, the perceiver senses the form of the object; first, by perceiving features of the object using our external senses. This sensory information is supplied to the internal senses, which merge all the pieces into a whole; a unified conscious experience. This process of perception and abstraction is the nexus of the soul and body. The material body may only perceive material objects, while the immaterial soul may only receive the immaterial; together, they form the universal. The soul and body interact in the final abstraction of the universal from; the concrete particular is the key to their relationship and interaction, which takes place in the physical body.

The soul completes the action of intellection by accepting forms that have been abstracted from matter. This process requires a concrete particular (material) to be abstracted into the universal intelligible (immaterial). The material and immaterial interact through the Active Intellect, which is a "divine light" containing the intelligible forms. The Active Intellect reveals the universals concealed in material objects much like the sun makes color available to our eyes.

Other contributions of Ibn Sina

- Astronomy and astrology

Ibn Sina's astronomical research and writings had influence on later researchers. One important feature of his work is that he considers mathematical astronomy as a separate discipline. He criticized Aristotle's view that the stars receiving their light from the Sun. Ibn Sina asserted that the stars are self-luminous. As evidence, he observed Venus as a spot on the Sun. This is possible, as there was a transit on 24 May 1032. He also used his transit observation establish that Venus was, at least sometimes, below the Sun in Ptolemaic cosmology, i.e. the sphere of Venus comes before the sphere of the Sun when moving out from the Earth in the prevailing geocentric model.

He wrote a treatise to correct the errors in Almagest, titled "to bring that which is stated in the Almagest and what is understood from Natural Science into conformity". For example, Ibn Sina considers the motion of the solar apogee, which Ptolemy had taken to be fixed.

Ibn Sina wrote an attack on astrology titled Resāla fī ebṭāl aḥkām al-nojūm, in which he cited passages from the Quran to dispute the power of astrology to foretell the future. He believed that each planet had some influence on the earth, but argued against astrologers being able to determine the exact effects.

- Chemistry

Ibn Sina was first to derive the attar of flowers from distillation and used steam distillation to produce essential oils such as rose essence, which he used as aromatherapeutic treatments for heart conditions.

Unlike al-Razi, Ibn Sina explicitly disputed the theory of the transmutation of substances commonly believed by alchemists:

Those of the chemical craft know well that no change can be affected in the different species of substances, though they can produce the appearance of such change.

Four works on alchemy attributed to Ibn Sina were translated into Latin as:

- Liber Aboali Abincine de Anima in arte Alchemiae
- Declaratio Lapis physici Ibn Sinae filio sui Aboali
- Ibn Sinae de congelatione et conglutinatione lapidum
- Ibn Sinae ad Hasan Regem epistola de Re recta

Liber Aboali Abincine de Anima in arte Alchemiae was the most influential, having influenced later chemists such as Vincent of Beauvais.

- Poetry

Much of Ibn Sina's works are versified. His poems appear in both Arabic and Persian.

- Middle Ages and Renaissance

Ibn Sina has been recognized by both East and West as one of the great figures in intellectual history. Johannes Kepler cites Ibn Sina's opinion when discussing the causes of planetary motions in Chapter 2 of Astronomia Nova. As early as the 14th century Dante Alighieri, in his Divine Comedy, depicted Ibn Sina in Limbo alongside the virtuous thinkers such as Virgil, Averroes, Homer, Horace, Ovid, Lucan, Socrates, Plato and Saladin.

George Sarton, the author of The History of Science, described Ibn Sina as "one of the greatest thinkers and medical scholars in history" and called him "the most famous scientist of Islam and one of the most famous of all races, places, and times". He was the world's leading writer in the field of medicine. His influence following translation of the Canon was such that from the early fourteenth to the mid-sixteenth centuries he was ranked with Hippocrates and Galen as one of the acknowledged authorities, princeps medicorum ("prince of physicians").

The Logic and Metaphysics have been extensively reprinted, the latter, e.g., at Venice in 1493, 1495 and 1546. Some of his shorter essays on medicine, logic, etc., take a poetical form (the poem on logic was published by Schmoelders in 1836). Two encyclopedic treatises, dealing with philosophy, are often mentioned. The larger, Al-Shifa' (Sanatio), exists nearly complete in manuscript in the Bodleian Library and elsewhere; part of it on the De Anima appeared at Pavia (1490) as the Liber Sextus Naturalium. There is also a حکمت مشرقیہ (hikmat-al-mashriqqiyya, in Latin Philosophia Orientalis), mentioned by Roger Bacon, the majority of which is lost in antiquity, which according to Averroes was pantheistic in tone.

Along with al-Razi, Al Zahrawi, Ibn al-Nafis and al-Ibadi, Ibn Sina is leading researcher of medicine. He is remembered in the Western history of medicine as a major historical figure who made important contributions to medicine and the European Renaissance.

List of works by Ibn Sina

The treatises of Ibn Sina influenced later Muslim thinkers in many areas including theology, philology, mathematics, astronomy, physics and music. His works numbered almost 450 volumes on a wide range of subjects, of which around 240 have survived. In particular, 150 volumes of his surviving works concentrate on philosophy and 40 of them concentrate on medicine. His most famous works are The Book of Healing, and The Canon of Medicine.

Following is a partial list of his works.

- Sirat al-shaykh al-ra'is (The Life of Ibn Sina), ed. and trans. WE. Gohlman, Albany, NY: State University of New York Press, 1974. The only critical edition of Ibn Sina's autobiography, supplemented with material from a biography by his student Abu 'Ubayd al-Juzjani. A more recent translation of the Autobiography appears in D. Gutas, Ibn Sina and the Aristotelian Tradition: Introduction to Reading Ibn Sina's Philosophical Works, Leiden: Brill, 1988; second edition 2014.

- Al-isharat wa al-tanbihat (Remarks and Admonitions), ed. S. Dunya, Cairo, 1960; parts translated by S.C. Inati, Remarks and Admonitions, Part One: Logic, Toronto, Ont.: Pontifical Institute for Medieval Studies, 1984, and Ibn Sina and Mysticism, Remarks and Admonitions: Part 4, London: Kegan Paul International, 1996.

- Al-Qanun fi'l-tibb (The Canon of Medicine), ed. I. a-Qashsh, Cairo, 1987. Encyclopedia of medicine manuscript, Latin translation, Flores Avicenne, Michael de Capella, 1508; Modern text, Ahmed Shawkat Al-Shatti, Jibran Jabbur.

- Risalah fi sirr al-qadar (Essay on the Secret of Destiny), trans. G. Hourani in Reason and Tradition in Islamic Ethics, Cambridge, Cambridge University Press, 1985.

- Danishnama-i 'ala'i (The Book of Scientific Knowledge), ed. and trans. P. Morewedge, "The Metaphysics of Ibn Sina", London, Routledge and Kegan Paul, 1973.

- Kitab al-Shifa' (The Book of Healing). Ibn Sina's major work on philosophy. He probably began to compose al-Shifa' in 1014, and completed it in 1020. Critical editions of the Arabic text have been published in Cairo, 1952–83, originally under the supervision of I. Madkour.

- Kitab al-Najat (The Book of Salvation), trans. F. Rahman, Ibn Sina's Psychology: An English Translation of Kitab al-Najat, Book II, Chapter VI with Historical-philosophical Notes and Textual Improvements on the Cairo Edition, Oxford: Oxford University Press, 1952. (The psychology of al-Shifa'.) (Digital version of the Arabic text).

- Risala fi'l-Ishq (A Treatise on Love). Translated by Emil L. Fackenheim.

- Ibn Sina's most important Persian work is the Danishnama-i 'Alai (دانشنامه علائی, "the Book of Knowledge for (Prince) 'Ala ad-Daulah". Ibn Sina created new scientific vocabulary that had not previously existed in Persian. The Danishnama covers such topics as logic, metaphysics, music theory and other sciences. It has been translated into English by Parwiz Morewedge in 1977. The book is also important in respect to Persian scientific works.

- Andar Danesh-e Rag (اندر دانش رگ, "On the Science of the Pulse") contains nine chapters on the science of the pulse and is a condensed synopsis.

- Persian poetry from Ibn Sina is recorded in various manuscripts and later anthologies such as Nozhat al-Majales.

Biographical Summary

Avicenna was born in c. 980 in the village of Afshana in Transoxiana to a Persian family. The village was near the Samanid capital of Bukhara, which was his mother's hometown. His father Abd Allah was a native of the city of Balkh in Tukharistan; he served as the governor in the royal estate of Harmaytan (near Bukhara) during the reign of Nuh II (r. 976–997).

Although both Ibn Sina's father and brother had adopted to Ismailism, he himself did not follow; and he continued as a Hanafi.

The family settled in Bukhara, a center of learning, which attracted many scholars. It was there that Ibn Sina was educated. He was first schooled in the Quran and literature, and by the age of 10, he had memorized the entire Quran. Afterwards, he was schooled in Jurisprudence by the Hanafi jurist Ismail al-Zahid. Sometime later,

Ibn Sina's father invited the physician and philosopher Abu Abdallah al-Natili to educate Ibn Sina. Thereafter Ibn Sina continued his research independently. Ibn Sina only mentions Natili as his teacher in his autobiography, but writers speculate about other teachers.

Ibn Sina started his career, at the age of 17, as a physician of Nuh II, amir of the Samanids (976–997). His father died when he was 21. Subsequently he became an administrator, possibly as the governor of Harmaytan.

Ibn Sina later moved to Gurganj, the capital of Khwarazm, where he served Ma'munid Abu al-Hasan Ali (r. 997 to 1009). The minister of Gurganj, Abu'l-Husayn as-Sahi was a patron of sciences. Under the Ma'munids, Gurganj became a centre of learning, attracting many prominent figures, such as Ibn Sina.

Ibn Sina moved in 1012, this time to the west. There he travelled through the Khurasani cities of Nasa, Abivard, Tus, Samangan and Jajarm. He was planning to visit the ruler of the city of Gurganj, the Ziyarid Qabus (r. 977–981, 997–1013), a cultivated patron of learning, whose court attracted many distinguished poets and scholars. However, when Ibn Sina eventually arrived, he discovered that the ruler had died in the winter of 1013. Ibn Sina then left Gurgan for Dihistan, but returned after becoming ill. There he had met Abu 'Ubayd al-Juzjani (died 1070) who became his pupil and companion. Ibn Sina stayed briefly in Gurganj, reportedly serving Qabus' son and successor Manuchihr (r. 1012–1031).

In c. 1014, Ibn Sina went to the city of Ray, where he entered into the service of the Buyid amir (ruler) Majd al-Dawla (r. 997–1029) and his mother Sayyida Shirin, the de facto ruler of the realm. There he served as the physician at the court, treating Majd al-Dawla, who was suffering from melancholia. Ibn Sina reportedly later served as the "business manager" of Sayyida Shirin in Qazvin and Hamadan. During this period, Ibn Sina finished his Canon of Medicine, and started writing his Book of Healing.

In 1015, during Ibn Sina's stay in Hamadan, he participated in a public debate, as was custom for newly arrived scholars in western Iran at that time. The debate helped to examine one's reputation against a prominent local resident. The person whom Ibn Sina debated against was Abu'l-Qasim al-Kirmani, a member of the school of philosophers of Baghdad. The debate became heated, resulting in Ibn Sina accusing Abu'l-Qasim of lack of basic knowledge in logic, while Abu'l-Qasim accused Ibn Sina of impoliteness. After the debate, Ibn Sina sent a letter to the Baghdad Peripatetics, asking if Abu'l-Qasim's claim that he shared the same opinion as them was true. Abu'l-Qasim later retaliated by writing a letter to an unknown person, in which he made accusations so serious, that Ibn Sina wrote to a deputy of Majd al-Dawla, named Abu Sa'd, to investigate the matter.

Not long afterwards, Ibn Sina shifted his allegiance to the rising Buyid amir Shams al-Dawla (the younger brother of Majd al-Dawla).

Ibn Sina had been invited by Shams al-Dawla to treat him, but subsequently asked Ibn Sina to become his vizier. Although Ibn Sina would sometimes clash with Shams al-Dawla's troops, he remained vizier until the latter died of colic (acute abdominal pain) in 1021. Ibn Sina was asked by Shams al-Dawla's son and successor Sama' al-Dawla (r. 1021–1023) to stay as vizier. However, Ibn Sina preferred to go with Abu Ghalib al-Attar, as his patron. It was during his stay at Attar's home that Ibn Sina completed his Book of Healing, writing 50 pages a day.

It was also during this period that Ibn Sina contacted Ala al-Dawla Muhammad (r. 1008–1041), the Kakuyid ruler of Isfahan and uncle of Sayyida Shirin. The Buyid court in Hamadan, particularly the Kurdish vizier Taj al-Mulk, suspected Ibn Sina of correspondence with Ala al-Dawla, and as a result had the house of Attar ransacked and Ibn Sina imprisoned in the fortress of Fardajan, outside Hamadan. Ibn Sina was imprisoned for four months, until Ala al-Dawla captured Hamadan, thus putting an end to Sama al-Dawla's reign.

Ibn Sina was released, and he went to Isfahan, where he was well received by Ala al-Dawla. In the words of Juzjani, the Kakuyid ruler gave Ibn Sina "the respect and esteem which someone like him deserved." Ibn Sina's service under Ala al-Dawla proved to be a stable period in his life. Ibn Sina served as the advisor, if not vizier of Ala al-Dawla, accompanying him in many of his military expeditions and travels. Ibn Sina dedicated two Persian works to him, a philosophical treatise named Danish-nama-yi Ala'i, and a medical treatise about the pulse.

During the brief occupation of Isfahan by the Ghaznavids in January 1030, Ibn Sina and Ala al-Dawla relocated to the southwestern Iranian region of Khuzistan, where they stayed until the death of the Ghaznavid ruler, Mahmud (r. 998–1030), two months later. When Ibn Sina returned to Isfahan, he started writing his "Pointers and Reminders". Ibn Sina had been suffering from the sickness of colic, an attack of acute abdominal pain localized in a hollow organ and often caused by spasm, obstruction, or twisting. In 1037, Ibn Sina accompanied Ala al-Dawla to a battle near Isfahan. While there, he had a severe case of colic and died shortly afterward in June 1037 AD in Hamadan, where he was buried.

Annotations

Ibn Sina wrote extensively on Islamic philosophy, especially on the subjects of logic, ethics and metaphysics. It included the treatises named "Logic" and "Metaphysics". Most of his works were written in Arabic, then the language of science. He wrote some in Persian. Of linguistic significance even to this day are a few books that he wrote in pure Persian language; for example, the Danishnamah-yi 'Ala'i' that he dedicated to Ala' ad-Dawla'.

Ibn Sina's commentaries on Aristotle often criticized the philosopher, encouraging a lively debate in the spirit of ijtihad (independent judgment).

Ibn Sina's theory of "emanations" became fundamental in the Kalam (school of theological discourse) in the 12th century. This theory is a metaphysical cosmology about the universe as an eternal, inevitable emanation or 'overflow' of the Divine, rather than as something God created out of nothing.

Islamic philosophy and Islamic metaphysics, imbued as it is with Islamic theology, clearly distinguishes between "essence" and "existence", more so than relatively fuzzy Aristotelianism. According to Ibn Sina, whereas existence is the domain of the contingent and the accidental, essence endures within a being beyond the accidental. This line of thought in philosophy and metaphysics, was already present in the works of al-Farabi who searched for a definitive Islamic philosophy separate from Occasionalism (a philosophical doctrine about causation which says that created substances cannot be efficient causes of events; instead, all events are taken to be caused directly by God). This can be seen in what is left of the works of al-Farabi. Following al-Farabi's lead, Ibn Sina initiated a full-fledged inquiry into the question of being, in which he distinguished between essence (Mahiat) and existence (Wujud).

Ibn Sina argued that the fact of existence cannot be inferred from or accounted for by the essence of existing things. Form and matter by themselves cannot interact and originate the movement of the universe, or the progressive actualization of existing things. Existence must, therefore, be due to an agent-cause that necessitates, imparts, gives, or adds existence to an essence.

To do so, the cause must be an existing thing and coexist with its effect.

This idea is foundational in metaphysics philosophy.

Ibn Sina's consideration of the essence-existence question may be elucidated in terms of his ontological analysis of the modalities of being. According to Ibn Sine beings are in three modes: impossibility, contingency and necessity. Impossibility mode of being is that which cannot exist; the contingent mode of being has the potentiality to be or not to be without entailing a contradiction; and a contingent mode can become actualized in to existence by an external cause settling between the choice to be or not to be. In contingent mode the existence is due to what is other than itself' (wajib al-wujud bi-ghayrihi). Thus, contingency-in-itself is potential beingness that could eventually be actualized by an external cause other than itself. The metaphysical structures of necessity and contingent modes are different. Necessity mode being due to itself (wajib al-wujud bi-dhatihi) is true in itself, while the contingent mode of being is 'false in itself' and 'true due to something else other than itself'. The necessity mode of being is the source of its own being without borrowed existence from an external cause; it is what always exists.

The necessity mode exists 'due-to-Its-Self', and has no quiddity or essence (mahiyya) other than existence (wujud). Furthermore, it is 'One' (wahid ahad) since there cannot be more than one 'Necessary-Existent-due-to-Itself' without differentia (fasl) to distinguish them from each other.

To require differentia entails that they exist 'due-to-themselves' (by hypothesis) as well as 'due to what is other than themselves' (because each can be the external cause for the other to become existent); and this is contradictory. However, if no differentia distinguishes them from each other, then there is no sense in which these 'Existents' are not one and the same.

Ibn Sina demonstrates, using similar arguments as above, that the 'Necessary-Existent-due-to-Itself' has no genus (jins), nor a definition (hadd), nor a counterpart (nadd), nor an opposite (did), and is detached (bari) from matter (madda), quality (kayf), quantity (kam), place (ayn), situation (wad) and time (waqt).

Ibn Sina's theology on metaphysical issues (ilāhiyyāt) has been criticized by some Islamic scholars, among them al-Ghazali, Ibn Taymiyya and Ibn al-Qayyim. While discussing the views of the theists among the Greek philosophers, namely Socrates, Plato and Aristotle in Al-Munqidh min ad-Dalal ("Deliverance from Error"), al-Ghazali noted that the Greek philosophers "must be taxed with unbelief, as must their partisans among the Muslim philosophers, such as Ibn Sina and al-Farabi and their likes." He added that "None, however, of the

Muslim philosophers engaged so much in transmitting Aristotle's lore as did the two men just mentioned. [...]

These critics state that "the sum of what we regard as the authentic philosophy by al-Farabi and Ibn Sina, can be reduced to three parts: a part which must be branded as unbelief; a part which must be stigmatized as innovation; and a part which need not be repudiated at all."

Some might regard such critics as populists versus logical.

Ibn Sina made an argument for the existence of God which would be known as the "Proof of the Truthful" (Arabic: burhan al-siddiqin). Ibn Sina argued that there must be a "necessary existent" (Arabic: wajib al-wujud), an entity that cannot not exist and through a series of arguments, he identified it with the Islamic conception of God. Ibn Sina's argument is one of the most influential arguments for God's existence, and it is plausibly Ibn Sina's biggest contribution to philosophy.

Correspondence between Ibn Sina with his student Ahmad ibn 'Ali al-Ma'sumi and Al-Biruni has survived in which they debated Aristotelian natural philosophy and the Peripatetic school, a philosophical school founded in 335 BC by Aristotle in the Lyceum in Ancient Athens. Abu Rayhan began by asking Ibn Sina eighteen questions, ten of which were criticisms of Aristotle's "On the Heavens".

Ibn Sina was a devout Muslim and sought to reconcile rational philosophy with Islamic theology. His aim was to prove the existence of God and His creation of the world scientifically and through reason and logic. Ibn Sina's views on Islamic theology (and philosophy) were enormously influential, forming part of the core of the curriculum at Islamic religious schools until the 19th century. Ibn Sina wrote a number of short treatises dealing with Islamic theology. These included treatises on the prophets (whom he viewed as "inspired philosophers"), and also on various scientific and philosophical interpretations of the Quran, such as how Quranic cosmology corresponds to his own philosophical system. In general, these treatises linked his philosophical writings to Islamic religious ideas; for example, the body's afterlife.

Some people claim that there are occasional brief hints and allusions in his longer works that Ibn Sina considered philosophy as the only sensible way to distinguish real prophecy from illusion. Such claimants further assert that he did not state this more clearly because of the impression of an implication of such a theory, that prophecy could be questioned. Such claimants further assert that most of the time Ibn Sina was busy writing shorter works which concentrated on explaining his theories on philosophy and theology clearly, without digressing to consider epistemological matters, which could be properly considered by other philosophers.

Later interpretations of Ibn Sina's philosophy split into three different schools; those (such as al-Tusi) who continued to apply his philosophy as a system to interpret later political events and scientific advances; those (such as al-Razi) who considered Ibn Sina's theological works in isolation from his wider philosophical concerns; and those (such as al-Ghazali) who selectively used parts of his philosophy to support their own attempts to gain greater spiritual insights through a variety of mystical means.

It was the theological interpretation championed by those such as al-Razi which eventually came to predominate.

Ibn Sina memorized the Quran by the age of ten, and as an adult, he wrote five treatises commenting on suras from the Quran. One of these texts included the Proof of Prophecies, in which he comments on several Quranic verses and holds the Quran in high esteem.

Ibn Sina argued that Islamic prophets are higher than philosophers.

Ibn Sina is generally understood to have been aligned with the Sunni Hanafi school of thought. He studied Hanafi law, many of his notable teachers were Hanafi jurists, and he served under the Hanafi court of Ali ibn Mamun. Ibn Sina said at an early age that he remained "unconvinced" by Ismaili missionary attempts to convert him. Historian Ẓahīr al-dīn al-Bayhaqī (d. 1169) also believed Ibn Sina to be a follower of the Brethren of Purity (إخوان الصفا); a secret society of Muslim philosophers in Basra, Iraq, in the 9th or 10th century AD.

Thought experiments: Floating man

While he was imprisoned in the castle of Fardajan near Hamadhan, Ibn Sina wrote his famous "floating man", a thought experiment to demonstrate human self-awareness and the substantiality and immateriality of the soul. This thought experiment demonstrated that the soul is a substance, and claimed humans cannot doubt their own consciousness, even in a situation that prevents all sensory data input. The thought experiment told its readers to imagine themselves created all at once while suspended in the air, isolated from all sensations, which includes no sensory contact with even their own bodies. He argued that, in this scenario, one would still have self-consciousness. Because it is conceivable that a person, suspended in air while cut off from sense experience, would still be capable of determining his own existence. The thought experiment points to the conclusions that the soul is a perfection, independent of the body, and an immaterial substance. The self-conceivability of this "Floating Man" indicates that the soul is perceived intellectually, which entails the soul's separateness from the body. Ibn Sina referred to the living human intelligence, particularly the Active Intellect, which he believed to be the hypostasis by which God communicates truth to the human mind that imparts order and intelligibility about nature. Following is an English translation of the argument:

One of us (i.e., a human being) should be imagined as having been created in a single stroke; created perfect and complete but with his vision obscured so that he cannot perceive external entities; created falling through air or a void, in such a manner that he is not struck by the firmness of the air in any way that compels him to feel it, and with his limbs separated so that they do not come in contact with or touch each other. Then contemplate the following: can he be assured of the existence of himself? He does not have any doubt in that his self exists, without thereby asserting that he has any exterior limbs, nor any internal organs, neither heart nor brain, nor any one of the exterior things at all; but rather he can affirm the existence of himself, without thereby asserting there that this self has any extension in space. Even if it were possible for him in that state to imagine a hand or any other limb, he would not imagine it as being a part of his self, nor as a condition for the existence of that self; for as you know that which is asserted is different from that which is not asserted and that which is inferred is different from that which is not inferred. Therefore, the self, the existence of which has been asserted, is a unique characteristic, in as much that it is not as such the same as the body or the limbs, which have not been ascertained. Thus, that which is ascertained (i.e., the self), does have a way of being sure of the existence of the soul as something other than the body, even something non-bodily; this he knows, this he should understand intuitively, if it is that he is ignorant of it and might need to be beaten with a stick [to realize it].

— Ibn Sina, Kitab Al-Shifa, On the Soul

However, Ibn Sina posited the brain as the place where reason interacts with sensation. Sensation prepares the soul to receive rational concepts from the universal Agent Intellect. The first knowledge of the flying person would be "I am," affirming his or her essence. That essence could not be the body, obviously, as the flying person has no sensation. Thus, the awareness that "I am" is the core of a human being: the soul exists and is self-aware. Ibn Sina thus concluded that the idea of the self is not logically dependent on any physical thing, and that the soul should not be seen in relative terms, but as a primary given, a substance. The body is unnecessary; in relation to it, the soul is its perfection. In itself, the soul is an immaterial substance.

21. Ibn Jazla

Abu Ali Yahya ibn Isa ibn Jazla al-Baghdadi

(Arabic: أبو علي يحيى بن عيسى بن جزله البغدادي),

was an 11th-century physician and author of an influential treatise on regimen.

Scientific Contributions

Ibn Jazla wrote a treatise titled "Taqwim al-Abdan fi Tadbir al-Insan (Dispositio corporum de constitutione hominis, Tacuin agritudinum), as the name implies it consists of tables in which diseases are arranged like the stars in astronomical tables.

The Tacuin was translated into Latin by Faraj ben Salim and the Latin version was published in 1532. A German translation was published at Strasbourg in 1533 by Hans Schotte.

Ibn Jazla also wrote another work, Al-Minhaj fi Al-Adwiah Al-Murakkabah, (Methodology of Compound Drugs), which was translated by Jambolinus and was known in Latin translation as the Cibis et medicines simplicibus.

A convert to Islam, he wrote works in praise of Islam and critical of Christianity and Judaism.

Biographical Summary

Ibn Jazla was born at Baghdad. He converted to Islam in 1074. He was under the tutelage of Abu `Ali ibn Al-Walid Al-Maghribi. He died in 1100.

22. Ibn al-Wafid

ʿAlī ibn al-Ḥusayn ibn al-Wāfid al-Lakhmī

(Arabic: علي بن الحسين بن الوافد اللخمي),

(c. 997 – 1074),

was a pharmacologist and physician from Toledo.

Scientific Contributions

Ibn al-Wafid's main work is Kitāb al-adwiya al-mufrada " كتاب الأدوية المفردة".

It was translated into Latin as De medicamentis simplicibus.

Ibn al-Wafid was mainly a pharmacist who used the techniques and methods available in alchemy to extract at least 520 different kinds of medicines from various plants and herbs.

His student Ali Ibn al-Lukuh was the author of ʿUmdat al-Ṭabīb fī Maʿrifat al-Nabāt li kulli Labīb, a famous botanical encyclopedia.

Biographical Summary

Ibn al-Wāfid was the vizier of Al-Mamun in Toledo. He is erroneously known in Europe as Abenguefith.

Ibn al-Wāfid was born in 997 AD and he died in 1074.

23. Ibn Butlan

Yawānīs al-Mukhtār ibn al-Ḥasan ibn ʿAbdūn al-Baghdādī,

also known as Ibn Buṭlān (Arabic: ابن بطلان),

(1001, Baghdad - 1064, Antioch),

was a physician in Baghdad.

Scientific Contributions

Ibn Buṭlān wrote the Taqwim al-Sihhah (تقويم الصحة); The Maintenance of Health. It treated matters of hygiene, dietetics, and exercise. It emphasized the benefits of regular attention to the personal physical and mental well-being.

Biographical Summary

Ibn Buṭlān was born in 1038 AD, presumably in Baghdad, and he died in 1075.

24. Ibn Zuhr

Abū Marwān 'Abd al-Malik ibn Zuhr

(Arabic: أبو مروان عبد الملك بن زهر),

(1094–1162),

was a physician, surgeon, and poet.

Scientific Contributions

Ibn Zuhr's major work was Al-Taysīr fil-Mudāwāt wal-Tadbīr ("Book of Simplification Concerning Therapeutics and Diet"). It was translated into Latin and Hebrew and greatly stimulated and accelerated the progress of surgery in Europe. He also improved surgical and medical knowledge by keying out several diseases and their treatments.

Ibn Zuhr performed the first experimental tracheotomy on a goat. He is thought to have made the earliest description of bezoar stones as medicinal items.

He was a contemporary of Ibn Rushd (Europeans erroneously call him Averroes) and Ibn Tufail, and was the most well-regarded physician of his era. He was particularly known for his emphasis on a more rational, empirical basis of medicine.

Kitab al-Iqtisad, "The book of moderation," was a treatise on general therapy written in Ibn Zuhr's youth for the Almoravid prince Ibrahim Yusuf ibn Tashfin. The book is a summary of various different diseases, therapeutics and general hygiene. It is also noted for its advice regarding cosmetics and physical beauty. Ibn Zuhr even recommended plastic surgery to alter acquired features such as big noses, thick lips or crooked teeth.

Kitab al-Aghdhiya, "The book of foods", is a manual on foods and regimen which contains guidelines for a healthy life. Ibn Zuhr wrote the book shortly after he went out of jail for his new patron, Almohad leader Abd al-Mu'min. The book contains classification of different kinds of dishes and foods like bread, meat, beverages, fruits and sweets. When he talks about the meat, Ibn Zuhr mentions different kinds of animals' fleshes, even unusual ones like those of gazelles, lions and snakes, classifying them based on their taste, usefulness and digestibility. He also recommends specific foods for each season of the year. For example, during winter, digestion is accelerated, so the amount of food consumed should also be increased. Moreover, the food should also be warmer and drier, as temperatures are lower and humidity is higher.

Kitab al-Taysir seems to be the last book of Ibn Zuhr before his death. As mentioned in the introduction, the book was authored at the request of his friend, Ibn Rushd, to act as a compendium to his medical encyclopedia "Colliget" which focused more on general topics of medicine. The two books were later translated into Hebrew and Latin, where they used to be printed as a single book and remained popular as late as the 18th century, and was part of the stimuli for European Renaissance.

The book, which contains 30 chapters, provides clinical descriptions and diagnosis of diseases starting from the head.

Ibn Zuhr provided an accurate description of the esophageal, stomach and mediastinal cancers, as well as other lesions. He proposed feeding enemas to keep alive patients with stomach cancer.

He was the first to give pathological descriptions of inflammations like otitis media and pericarditis.

Ibn Zuhr provided one of the earliest recorded evidence of the Scabies mite, which contributed to the scientific advancement of microbiology. In his Kitab al-Taysir, he wrote the following:

There are lice under the hand, ankle and foot, like worms, and sores affecting the same areas. If the skin is removed, there appears from various parts of it, a very small animal which can hardly be seen.

Ibn Zuhr's greatest contribution, not only to medicine but to the culture of science in general, is his emphasis on experimental approach. For example, he introduced testing and experimenting using animals. He performed medical procedures on animals before doing them on humans to know if they would work. Most notable was his approval and recommendation for the surgical procedure of tracheotomy, which was a controversial procedure at the time. In trying to sort out the controversy, Ibn Zuhr described the following medical experiment which he performed on a goat:

"Earlier on in my training when I read those opinions (controversies), I cut on the lung pipe of a goat after incising the skin and the covering sheath underneath. Then I completely cut off the substance of the pipe, an area just less than the size of a tirmisah (lupine seed). Then, I kept washing the wound with water and honey till it healed and it (the animal) totally recovered and lived for a long time."

Ibn Zuhr's contribution to the culture of science in general through his emphasis on experimental approach. For example, according to Aristotle, whose writings had remained unquestioned, not only did heavier objects fall faster than lighter ones, but an object that weighed twice as much as another would fall twice as fast. According to the emphasis on experimentation in the Muslim world, instead of blindly believing Aristotle about the falling bodies, or arguing philosophically about it, the preferred way is to do experimentation. Just drop the falling objects from the leaning tower of Pisa and observe what happens. That is what Galileo actually did, thanks to the culture of science in the Muslim world which had been assimilated into Europe through the translations of the works of the Muslim Scientists.

Ibn Abi Usaibia mentions these other works of Ibn Zuhr:

- Fi al-Zinah (On Beatification).

- Al-Tiryaq al-Sabini (On Antidotes).

- Fi Illat al-Kila (On Diseases of the Kidney).

- Fi Illat al-Baras wa al-Bahaq (On Leprosy and Vitiligo).

- Al-Tadhkirah (The Remembrance).

Bibliographical Summary

Ibn Zuhr was born in 1094 AD at Seville, Andalusia (present-day Spain). He was from the notable Banu Zuhr family who were members of the Arab tribe of Iyad. The family had produced, since the early 10th century, six consecutive generations of physicians; and also included jurists, poets, viziers, courtiers, and midwives who served under the rulers of Andalusia.

Ibn Zuhr started his education by studying religion and literature, as was the custom. He later studied medicine with his father, Abu'l-Ala Zuhr (d.1131) at an early age.

Ibn Zuhr started his medical career as a court physician for the Almoravid empire. However, he later fell out of favor with the Almoravid ruler, 'Ali bin Yusuf bin Tashufin, and fled from Seville. He was however, apprehended and jailed in Marrakesh in 1140. Later in 1147 when the Almohad empire conquered Seville, he returned and devoted himself to medical practice.

According to Leo Africanus, ibn Zuhr attended lectures of Ibn Rushd, and learned physic from him.

Ibn Zuhr died in Seville in 1162.

25. Al-Ghafiqi

Muhammad ibn Aslam Al-Ghafiqi

(Arabic: محمد بن أسلم الغافقي),

(d. 1165 CE),

was a 12th-century Ophthalmologist from Andalusia.

Scientific Contributions

Al-Ghafiqi wrote a treatise: Al-Murshid fi 'l-Kuhhl (The Right Guide to Ophthalmology). The book shows a deep understanding of the conditions of the eye and eyelids, and their successful treatment with many different surgical procedures, ointments, and chemical medicines.

Biographical Summary

Al-Ghafiqi d. 1165 CE) was an Andalusian ophthalmologist. He died in 1165 AD.

26. Abu al-Majd ibn Abi al-Hakam

Abu al-Majd ibn Abi al-Hakam Ubaydullah Ibn al-Muzaffar al-Bahili

(Arabic: أفضل الدولة أبو المجد محمد بن أبي الحكم عبيد الله بن المظفر بن عبد الله الباهلي),

(d. 1174 CE),

was an Andalusian physician, musician and astrologer from Damascus, Syria.

Scientific Contributions

When Nur ad-Din Zangi founded the Bimaristan in Damascus, he entrusted the medical care of the patients to Abu al-Majd ibn Abi al-Hakam. Of Ibn Abi al-Hakam, the historian Ibn Abi Usaybi'a states:

> He used to examine the patients at the hospital, take note of their condition and listen to their complaints in the presence of the nurses and porters, who were charged with looking after them. All methods of treatment and prescriptions which he gave were carried out to the letter. After finishing his ward round, he would pass on to a magnificently furnished hall to consult various scientific works; and all physicians and students gathered round him in order to discuss medical points. When he had asked questions of his pupils and worked for about three hours in the wards and in the library, he would return to his home".

105

Abu al-Majd ibn Abi al-Hakam invented the Bimaristan (the place of the sick) as a fully staffed and equipped hospital. He ran it as a full function teaching hospital, (not just as a practical demonstration piece of it, as it happens today).

Biographical Summary

Abu al-Majd ibn Abi al-Hakam lived in Damascus, Syria. He ran his hospital as a full-function teaching facility. He died in 1174 AD.

27. Ibn Hubal

Muhadhdhib al-Dīn Abū'l-Hasan 'Alī ibn Ahmad Ibn Hubal

(Arabic: مهذب الدين أبي الحس علي بن أحمد ابن هبل).

also known as Ibn Hubal (Arabic: ابن هبل),

(c. 1122 - 1213),

was a physician and scientist born in Baghdad.

Scientific Contributions

Ibn Hubal was known primarily for his medical treatise titled "Kitab al-Mukhtarat fi al-tibb (Arabic: كتاب المختارات في الطب)". It was written in 1165 in Mosul, north of Baghdad, where Ibn Hubal spent most of his life.

The chapters on kidney and bladder stones were translated into French by P. de Koning in his Traité sur le calcul dans les reins et dans la vessie (1896).

Other chapters have been translated by Dorothee Thies in Die Lehren der arabischen Mediziner Tabari und Ibn Hubal über Herz, Lunge, Gallenblase und Milz (1968).

Biographical Summary

Ibn Hubal lived much of his life in Mosul, Iraq. He was born in 1122 AD and he died in 1213.

28. Al-Dakhwar

Muhadhdhabuddin Abd al-Rahim bin Ali bin Hamid al-Dimashqi

(Arabic: مهذب الدين عبد الرحيم بن علي بن حامد الدمشقي),

also known as al-Dakhwar (Arabic: الدخوار)

(1170–1230),

was a leading physician who served many rulers of the Ayyubid dynasty.

Scientific Contributions

Al-Dakhwar was also responsible for medicine in Cairo and Damascus. Al-Dakhwar educated or influenced most of the prominent physicians of Egypt and Syria, including Ibn al-Nafis, the discoverer of blood circulation in the human body.

In 1208, al-Adil, the Sultan of Egypt, told his vizier al-Sahib ibn Shukur, that he was in need of another physician with the equivalent skill of the chief of medicine at the time, Abd al-Aziz al-Sulami. Al-Adil believed that al-Sulami was busy enough serving as physician of the army. Ibn Shukur recommended al-Dakhwar for the post and offered him 30 dinars a month. Al-Dakhwar turned him down, citing that al-Sulami receives 100 dinars a month and stating "I know my ability in this field and I will not take less!" Al-Sulami died on June 7 and soon after al-Dakhwar himself came into contact with al-Adil, and the latter was greatly impressed by him. He not only appointed him as his personal physician, but also as one of his confidants.

When al-Adil died, his son and successor in Damascus, al-Mu'azzam, made him chief superintendent of the Nasiri Hospital.

There he gave lectures on medicine, and wrote books. Following are some of his books on medicine.

- al-Janinah ("The Embryo")
- Sharh Taqdimat-il-Ma'rifah ("Commentary on the Introduction of Knowledge")
- Mukhtasar-ul-Hawl-il-Razi ("Resume of al-Hawi by al-Razi")

He also wrote a book on poetry: Kitab ul-Aghani (a summarized version of "The Book of Songs" by al-Isfahani).

Later, when al-Adil's other son al-Ashraf annexed Damascus after al-Mu'azzam died, al-Dakhwar was promoted as chief medical officer of the Ayyubid state.

Biographical Summary

Al-Dakhwar was born and brought up in Damascus in 1170 AD. His father was an ophthalmologist. He too was an ophthalmologist at the Nuri Hospital of Damascus.

Later, he further studied medicine with Ibn al-Matran.

Al-Dakhwar died in 1230 AD.

29. Abd al-Latif al-Baghdadi

Muwaffaq al-Dīn Muḥammad ʿAbd al-Laṭīf ibn Yūsuf al-Baghdādī
(Arabic: موفق الدين محمد عبد اللطيف بن يوسف البغدادي),

(1162 Baghdad–1231 Baghdad),

was a physician, philosopher, historian, Arabic grammarian and
traveler, and one of the most voluminous writers of his time.

Scientific Contributions

ʿAbd al-Laṭīf was a man of great knowledge and of an inquisitive and
penetrating mind. He wrote numerous works (mostly on medicine).
Ibn Abī Uṣaybiʿah ascribes to him numerous of these works.
However, Europeans seem to know him for his graphic and detailed
Account of Egypt in two parts.

In Archeology, ʿAbd-al-Laṭīf was well aware of the value of the
ancient monuments. He praised some Muslim rulers for preserving
and protecting artefacts and monuments, and criticized others for
failing to do so. He noted that the preservation of antiquities
presented a number of benefits for Muslims because "monuments are
useful historical evidence for chronologies" and "they furnish evidence
for Holy Scriptures, since the Qur'an mentions them and their
people"; further, "they are reminders of human endurance and fate"
and "they show, to a degree, the politics and history of ancestors, the
richness of their sciences, and the genius of their thought".

While discussing the profession of treasure hunting, he notes that poorer treasure hunters were often sponsored by rich businessmen to go on archeological expeditions. In some cases, an expedition could turn out to be fraudulent, with the treasure hunter disappearing with large amounts of money extracted from sponsors.

In Egyptology, his manuscript was one of the earliest works. It contains a vivid description of a famine caused by the Nile failing to overflow its banks; which occurred during the author's residence in Egypt. He also wrote detailed descriptions on ancient Egyptian monuments.

In the medical science of Autopsy, he wrote that during the famine in Egypt in 597 AH (1200 AD), he had the opportunity to observe and examine a large number of skeletons, through which he came to discover that Galen was in error regarding the formation of the bones of the lower jaw [mandible], coccyx and sacrum.

As far as philosophy is concerned, one may adduce that ʿAbd al-Laṭīf al-Baghdādī regarded philosophers as paragons of real virtue and therefore he refused to accept as a true philosopher one lacking not only true insight, but also a truly moral personality, as true philosophy was in the service of religion, verifying both belief and action.

ʿAbd al-Laṭīf composed several philosophical works, among which is an important and original commentary on Metaphysics, (Kitāb fī ʿilm mā baʿd al-ṭabīʿa). This work is in continuity with the writings of al-Kindī (d. circa 185-252/801-66) and al-Fārābī (d. 339/950).

The philosophical section of his Book of the Two Pieces of Advice (Kitāb al-Naṣīḥatayn) contains an interesting and challenging defense of philosophy and illustrates the vibrancy of philosophical debate in the Islamic society. It moreover emphasizes that, contrary to the commonly prevailing idea, Islamic philosophy did not decline after the twelfth century CE.

In Alchemy, ʿAbd al-Laṭīf penned two passionate pamphlets against the art of alchemy in all its facets. Although he engaged in alchemy for a short while, he later abandoned the art completely by rejecting not only its practice, but also its theory. In his view alchemy could not be placed in the system of the sciences, and its false presumptions and pretensions must be distinguished from true scientific knowledge, which can be given a rational basis.

Al-Baghdādī's Arabic manuscript was discovered in 1665 and is preserved in the Bodleian Library. Pococke published the Arabic manuscript in the 1680s. His son, Edward Pococke the Younger, translated the work into Latin, although he was only able to publish less than half of this work. Thomas Hunt attempted to publish Pococke's complete translation in 1746, although his attempt was unsuccessful. Pococke's complete Latin translation was eventually published by Joseph White of Oxford in 1800. The work was then translated into French by Silvestre de Sacy in 1810.

Biographical Summary

Many details of 'Abd al-Laṭīf al-Baghdādī's life are known from Ibn Abī Uṣaybi'ah's literary history of medicine.

As a young man, 'Abd al-Laṭīf studied grammar, law, tradition, medicine, alchemy and philosophy. He adopted Ibn Sīnā as his philosophical mentor at the suggestion of a wandering scholar from the Maghreb. He travelled extensively and resided in Mosul (in 1189) where he studied the works of al-Suhrawardi before travelling on to Damascus (1190) and the camp of Saladin outside Acre (1191). It was at this last location that he met Baha al-Din ibn Shaddad and Imad al-Din al-Isfahani and acquired the Qadi al-Fadil's patronage. He went on to Cairo, where he met Abu'l-Qasim al-Shari'i, who introduced him to the works of al-Farabi.

In 1192 he met Saladin in Jerusalem and enjoyed his patronage, then went to Damascus again before returning to Cairo. He journeyed to Jerusalem and to Damascus in 1207-8, and eventually made his way via Aleppo to Erzindjan, where he remained at the court of the Mengujekid Ala'-al-Din Da'ud (Dāwūd Shāh) until the city was conquered by the Rūm Seljuk ruler Kayqubād II (Kayqubād Ibn Kaykhusraw). 'Abd al-Latif returned to Baghdad in 1229, travelling back via Erzerum, Kamakh, Divriği and Malatya. He died in Baghdad two years later in 1231 AD.

30. Rashidun al-Suri

Rashidun al-Suri

(Arabic: رشيد الدين الصوري),

(1177–1241),

was a leading physician and botanist in the 13th century, during the Ayyubid dynasty.

Scientific Contributions

Al-Suri met and greatly impressed the Ayyubid sultan al-Adil in the early 13th century. Al-Adil brought al-Suri to Cairo and made him his personal physician.

He also served al-Adil's son, al-Mu'azzam and grandson, an-Nasir Dawud, the successive governors of Damascus.

As part of his interest in medicine, al-Suri held an interest for plant life and was a botany researcher. He used to roam about and study herbs and plants in their natural habitat surroundings. This included the role of these plants as sources of medicine. He employed a professional painter to sketch and paint for him the plants in different stages of their growth, as accurate as possible, using various colors and dyes.

His book, entitled "Kitab al-Adwiyat al-Mufradah" ("The Simple Medicines") is not extant.

Biographical Summary

Al-Suri was born in 1177 in Tyre, then part of Lebanon and derives his name al-Suri from the name of the city in Arabic "Sur". After completing his preliminary education in Tyre, he moved to Jerusalem, under Ayyubid control, where he served as a physician at a hospital.

He died in 1241 AD.

31. Ibn al-Baitar

Diyā' al-Dīn Abū Muḥammad ʿAbd Allāh ibn Aḥmad al-Mālaqī, commonly known as Ibn al-Bayṭār

(Arabic: ابن البيطار),

(1197–1248 AD),

was a physician, botanist, and pharmacist from Andalusia.

Scientific Contributions

Ibn al-Bayṭār systematically recorded the additions to the list of medicines made by Muslim physicians: around 300 and 400 types of medicine were added. Thus, the total list was enlarged by about 50%, with the new medicines treating both the old diagnosis as well as the new diagnosis discovered by Ibn al-Bayṭār.

Kitāb al-Jāmiʿ li-Mufradāt al-Adwiya wa-l-Aghdhiya (Arabic: كتاب الجامع لمفردات الأدوية والأغذية) is Ibn al-Bayṭār's largest and most widely read book. It is his Compendium on Simple Medicaments and Foods. It is a pharmacopoeia (pharmaceutical encyclopedia) listing 1400 plants, foods, and drugs, and their uses. It is organized alphabetically by the name of the useful plant or plant component or other substance – a small minority of the items covered are not botanicals. For each item, Ibn al-Bayṭār makes one or two brief remarks himself and gives brief extracts from a handful of different authors. The book contains references to 150 previous Muslim authors, as well as 20 previous Greek authors.

He cites Book Two of the Canon of Medicine of Ibn Sīnā (Aveicenna). Ibn al-Bayṭār's treatments are richer in detail, and a large number of Ibn al-Bayṭār's useful plants or plant substances are not covered at all by earlier sources. In modern printed edition, the book is more than 900 pages long. It was published in full in Arabic. Its German and French translations were published in the 19th century.

Ibn al-Bayṭār provides detailed chemical information on the Rosewater and Orange water production. He mentions: The scented Shurub (Syrup) was often extracted from flowers and rare leaves, by means of using hot oils and fat; they were later cooled in cinnamon oil. The oils used were also extracted from sesame and olives. Essential oil was produced by joining various retorts; the steam from these retorts were condensed and combined; and its scented droplets were used as perfume; and mixed to produce the medicines.

Ibn al-Bayṭār's second major work is Kitāb al-Mughnī fī al-Adwiya al-Mufrada, an encyclopedia of medicine which incorporates his knowledge of plants used extensively for the treatment of various ailments, including diseases related to the head, ear, eye, etc.

Following are some other works of Ibn al-Bayṭār.

- Mīzān al-Ṭabīb.
- Risāla fī l-Aghdhiya wa-l-Adwiya.

- Maqāla fī al-Laymūn, Treatise on the Lemon (also attributed to Ibn Jumayʿ); translated into Latin by Andrea Alpago as Ebn Bitar de malis limonis (Venice 1593).
- Tafsīr Kitāb Diyāsqūrīdūs, a commentary on the first four books of Dioscorides' "Materia Medica."

Biographical Summary

Ibn al-Baitar was born in 1197 in the city of Málaga in al-Andalus (Muslim Spain) at the end of the twelfth century; hence his nisba "al-Mālaqī". His name "Ibn al-Baitar" is Arabic for "son of the veterinarian", which was his father's profession.

Ibn al-Bayṭār learned botany from the Málagan botanist Abū al-ʿAbbās al-Nabātī with whom he started collecting plants in and around Spain.

Al-Nabātī had developed a scientific method, that invented empirical and experimental techniques to test, describe, and identify numerous materia medica. Using this science, he separated unverified reports from those supported by actual tests and observations.

This approach was adopted by Ibn al-Bayṭār.

In 1219, Ibn al-Bayṭār left Málaga, travelling to the coast of North Africa and as far as Anatolia, to collect plants. He researched the plants in Marrakech, Bugia, Constantinople, Tunis, Tripoli, Barqa and Antalya.

After 1224, he entered the service of the Ayyubid Sultan al-Kāmil and was appointed chief herbalist. In 1227 al-Kāmil extended his domination to Damascus, and Ibn al-Bayṭār accompanied him there, which provided him an opportunity to research plants in Syria. His botanical researches extended over a vast area including Arabia and Palestine.

Ibn al-Bayṭār used the term "snow of China" (in Arabic, thalj al-Ṣīn) to describe saltpetre while writing about gunpowder.

Ibn al-Baitar died in Damascus in 1248 AD.

32. Ibn Abi Usaybi'a

Ibn Abī Uṣaybiʿa Muʾaffaq al-Dīn Abū al-ʿAbbās Aḥmad Ibn Al-Qāsim Ibn Khalīfa al-Khazrajī

commonly Ibn Abi Usaibia,

(also Usaibi'ah, Usaybea, Usaibi`a, Usaybiʿah, etc.),

(Arabic: ابن أبي أصيبعة),

(1203–1270),

was a physician from Syria in the 13th century CE. He compiled a biographical encyclopedia of notable physicians.

Scientific Contributions

The title in Arabic, Uyūn ul-Anbāʾ fī Ṭabaqāt al-Aṭibbā (Arabic: عيون الأنباء في طبقات الأطباء), is translatable loosely and expansively as "Sources of News on Classes of Physicians", commonly translated into English as History of Physicians, Lives of the Physicians, Classes of Physicians, or Biographical Encyclopedia of Physicians. The book opens with a summary of the physicians from ancient Greece, Syria, India and Rome but the main focus of the book's 700 pages is Muslim physicians. A first version appeared in 1245–1246 and was dedicated to the Ayyubid physician and vizier Amīn al-Dawlah. A second and enlarged recension of the work was produced in the last years of the life of the author; the extant manuscripts have multiple versions presumably due to the haphazard translation processes used by the Europeans.

In current times, the text has been published seven times. When the first edition by August Müller (Cairo, 1882) was published under the pseudonym "Imrū l-Qays", it was found to be marred by typos and errors. Hence a corrected version was subsequently issued (Königsberg, 1884). Relying on Müller's work, Niẓār Riḍā published an edition of the text in Beirut in 1965. Beirut edition was subsequently reworked by Qāsim Wahhāb for another edition issued in Beirut in 1997. ʿĀmir al-Najjār published his own critical edition in Cairo in 1996, that was not based on Müller's editions.

A team from the universities of Oxford and Warwick has published a new critical edition and a full annotated English translation of the Uyūn al-Anbā. Their work is available in Open Access at Brill Scholarly Editions.

In 2020, a new translation was published by Oxford World's Classics under the name Anecdotes and Antidotes: A Medieval Arabic History of Physicians. This title somewhat belittles the scholarship in the original Arabic Text.

Biographical Summary

Ibn Abi Usaibia was born at Damascus, a member of the Banu Khazraj tribe. The son of a physician, he studied medicine at Damascus and Cairo. In 1236, he was appointed physician to a new hospital in Cairo, but the following year he took up an offer by ruler of Damascus, of a post in Salkhad, near Damascus, where he lived until his death. His only surviving work is the Uyūn al-Anbā. In that work, he mentions another of his works, but it has not survived.

33. Ibn al-Quff

Amīn-ad-Daula Abu-'l-Faraǧ ibn Ya'qūb ibn Isḥāq Ibn al-Quff al-Karaki

(Arabic: أمين الدولة أبو الفرج بن يعقوب بن إسحاق بن القف الكركي),

AD 1233–1286),

was a physician and surgeon.

Scientific Contributions

During his time in Jordan, as a physician-surgeon, Ibn al-Quff was writing his books, and teaching. He was actually more well known as a writer and educator on medical topics than as a doctor. He wrote ten commentaries and books during his lifetime. Seven of these works are known to exist today whether fragments or the entire work. One of his allegedly most famous works was a commentary on Ishārāt of Ibn Sīnā but there is no evidence of this today; it went missing, or Ibn al-Quff never finished it.

Some of his surviving works are listed below.

- Kitāb al-'Umda fi 'l-ǧirāḥa (كتاب العمدة في الجراحة) or *Basics in the Art of* Surgery: The work was published in Hyderabad, India in 1937.

- Al-Shafi al-Tibb (the Healing Arts of medicine): completed early 1272 AD.

- *Jāmiʿ al-gharaḍ fī ḥifẓ al-ṣiḥḥah wa-dafʿ al-maraḍ* (جامع الغرض في حفظ الصحة ودفع المرض): on preventive medicine and the preservation of health. It has 60 chapters, and was completed around 1275 AD. It is extant in several manuscripts.

- *Al-usul fi sarh al-fusul*: A two-volume commentary of the works of Hippocrates.

- *Risala fi manafi al-a da*: A treatise on the anatomy of the body's organs.

- *Zubad at-Tabib*: A book with advice for practicing physicians.

- *Sarh al-Kulliyat*: A commentary on Ibn Sina's work *Qanun fit-Tibb*.

Biographical Summary

Ibn al-Quff, was born in 1233 AD in the city of Al Karak (in modern-day Jordan). The family moved to Sarkhad in Syria, where Ibn al-Quff was tutored by Ibn Abi Uṣaybiʿah.

Ibn al-Quff moved to Damascus where he improved his knowledge and studied metaphysics, philosophy, medicine, natural sciences, and mathematics.

Ibn al-Quff got the job of physician-surgeon in the army which was stationed in Jordan.

After his popularity died down, he was sent to Damascus and remained there teaching until his death in 1286 AD.

34. Ibn al-Nafis

Ala-al-Din abu al-Hasan Ali ibn Abi-Hazm al-Qarshi al-Dimashqi

(Arabic: علاء الدين أبو الحسن عليّ بن أبي حزم القرشي الدمشقي),

also known as Ibn al-Nafis

(Arabic: ابن النفيس),

was a polymath whose areas of work included medicine, surgery, physiology, anatomy, biology, Islamic studies, jurisprudence, and philosophy.

He is known for being the first to describe the pulmonary circulation of the blood.

Scientific Contributions

The work of Ibn al-Nafis regarding the right sided (pulmonary) circulation pre-dates the later work (1628) of William Harvey's De motu cordis.

2nd century Greek physician Galen's theory about the physiology of the circulatory system remained unchallenged until the revolutionary works of Ibn al-Nafis. He is "the father of circulatory physiology". However, European science community does not give acknowledgement to Ibn al-Nafis, recognizing instead William Harvey's De motu cordis. This is systemic dishonesty and lack of reliability within the European science tradition with respect to the contributions of the Muslim scientists.

As an early anatomist, Ibn al-Nafis also performed several human dissections during the course of his work, making several breakthrough discoveries in the fields of physiology and anatomy. Besides his famous discovery of the pulmonary circulation, he also gave an early insight of the coronary and capillary circulations. The number of medical textbooks written by Ibn al-Nafis is estimated at more than 110 volumes.

He was appointed as the chief physician at al-Naseri Hospital founded by Sultan Saladin.

Apart from medicine, Ibn al-Nafis studied jurisprudence, literature and theology. He was an expert on the Shafi'i school of jurisprudence.

Bibliographical Summary

Ibn al-Nafis was born in 1213 AD at a village near Damascus named Karashia. Early in his life, he studied theology, philosophy and literature. Then, at the age of 16, he started studying medicine for more than ten years at the Nuri Hospital in Damascus, which was founded by the Turkish Prince Nur-al Din Muhmud ibn Zanki, in the 12th century. He was contemporary with the famous Damascene physician Ibn Abi Usaibia and they both were taught by the founder of a medical school in Damascus, Al-Dakhwar. Ibn Abi Usaibia does not mention Ibn al-Nafis at all in his biographical dictionary "Lives of the Physicians". The seemingly intentional omission could be due to rivalry between the two physicians.

In 1236, Ibn al-Nafis, along with some of his colleagues, moved to Egypt under the request of the Ayyubid sultan al-Kamil. Ibn al-Nafis was appointed as the chief physician at al-Naseri hospital which was founded by Saladin. There, Ibn al-Nafis taught and practiced medicine for several years. Ibn al-Nafis also taught jurisprudence at al-Masruriyya Madrassa (Arabic: المدرسة المسرورية). His name is found among those of other scholars, which gives insight into how well he was regarded in the study and practice of jurisprudence.

Ibn al-Nafis lived most of his life in Egypt, and witnessed several pivotal events like the fall of Baghdad and the rise of Mamluks. He even became the personal physician of the sultan Baibars and other prominent political leaders, thus becoming an authority among practitioners of medicine. Later in his life, when he was 74 years old, Ibn al-Nafis was appointed as the chief physician of the newly founded al-Mansori hospital where he worked for the rest of his life.

Ibn al-Nafis died in 1288 AD in Cairo, after a brief sickness. Prior to his death, he donated his house and library to Qalawun Hospital.

35. Al-Suwaydi

'Izz al-Dīn Abū Isḥāq Ibrāhīm ibn Muḥammad ibn Ṭarkhān as-Suwaydī

(Arabic: ابراهيم ابن محمد ابن طرخان السويدى),

(1204–1292),

was a physician.

Scientific Contributions

As-Suwaydi was active in Cairo and Damascus.

As-Suwaydi's treatise Tadhkirah was epitomized by Sha'rānī in the 16th century. He wrote three works:

a treatise on plant names,

a treatise on the medical use of stones, and

a book of medical recipes and procedures (Tadhkirah).

Biographical Summary

As-Suwaydi was born in 1204 AD, and he died in 1292. He was from the Aws tribe.

As-Suwaydi was a pupil of Ibn al-Baytar.

36. Ibn al-Akfani

Muhammad ibn Ibrāhīm ibn al-Akfani

(Arabic: ابن الأكفاني),

(1286-ca. 1348–49),

was a physician and an encyclopedist.

Scientific Contributions

Ibn al-Akfani wrote at least 22 books. Most of his books were science related, including logic, gemology, mathematics, medicine and astronomy. Specific subjects include bloodletting, slavery and ophthalmology.

His most famous work was a science encyclopedia called Iršād al-qāsid ilā asnā' al-maqāsid. The encyclopedia examines 60 subjects with bibliographies and a glossary of terms. His book, Kitāb nuhab al-dahā'ir fī ahwāl al-jawāhir, is about gemstones, with a focus about jacinth.

Biographical Summary

Ibn al-Akfani was born in 1286 AD in Sinjar, Iraq; and he died in 1348 or 1349 AD during the bubonic plague.

He lived in Cairo, Egypt, and worked at Al-Mansuri Hospital.

Concluding Remarks

We have presented 36 scientists in the Medical Sciences from the part 1 (AD 610 to 1400) of the Islamic Era (AD 610 to 1922). All of them except one are Muslims, an expression of the fact that the era was entirely dominated by the Muslims in all domains of medical sciences.

The series makes it explicit that there is a natural affinity between Islam and science because Quran exhorts its readers to a scientific outlook in life by urging them to observe the nature and the universe around them to get to know them. That is the solidly open path to appreciate the truth in Quran and to approach closer to its Speaker.

On the pragmatic aspect, the requirements of religious acts make it necessary to do scientific research in order to meet those requirements. Following are some examples:

- the need to pray five times a day necessitated reliable time keeping giving rise to various sciences of sun dials, researching for sciences of projection to produce equal shadow during each hour. Masajid employed the leading-edge astronomers as time keepers, and further encouraged them towards breakthroughs in sciences;

- the need to determine the direction of Qibla, over the vast Islamic empire, necessitated research in geography, astronomy, spherical geometry, trigonometry, spherical trigonometry, and analytic geometry (combining geometry and algebra);

- the requirement of wudu necessitated water management and aesthetic fountains in the courtyards of magnificent masajid;

- the requirement of Miqat in haj and umrah have necessitated geographical measurements and geometry calculations;

- the spiritual practices of haj and umrah had brought together Muslims into a world-wide congress, including for the scientific research conferencing;

- and the descriptions of Jannah inspired aesthetic sciences of gardening, water fountains, water-powered-clocks, water mills, and other technological marvels.

The natural synergy between Islam and sciences is inherently necessary, complementary, and mutually supportive. This synergy enabled the following absolutely foundational research contributions to the world civilization; these are some illustrative examples given in chronological order.

1. Let us start with Al-Khwarizmi who contributed the complete algorithms for decimal mathematics at a time when the Europeans were happily busy with their abacus. He calculated the circumference of the Earth, and developed an accurate world map. He wrote the treatise Kitāb al-mukhtaṣar fī ḥisāb al-jabr wal-muqābala (Arabic: الكتاب المختصر في حساب الجبر والمقابلة) which among other topics discusses "Algebra", details of its operations, solution of polynomial and quadratic equations, with applications in trade, surveying, and legal

inheritance. Invention of Algebra allowed mathematics to be applied to itself in a way which nobody had dreamed of, and was an unmistakable revolution in mathematics. The Greeks had not dreamed of such possibilities in mathematics.

2. Al-Jahiz[1] clearly discussed the struggle for existence and the determining factors of natural selection. He also discussed micro evolution. However, the theory of natural selection and evolution is attributed entirely to Darwin.

3. Al-Kindi did the earliest known use of statistical inference and wrote a book on cryptography and cryptanalysis. He invented methods of breaking ciphers using frequency analysis. However, encyclopedia Britannica in its current article on "History of Cryptology" does not even mention Al-Kindi.

4. Ahmad Ibn Yusuf researched "Ratios and Proportions" and worked sequence of calculations. However, the sequence got attributed entirely to Fibonacci.

5. Abu Kamil expanded Al-Khwarizmi's Algebra by introducing irrational numbers into mathematics; solved 8th order equations, equations with irrational numbers, and equations with integer solutions. He also solved sets of non-linear simultaneous equations with three unknown variables. Abu

---

[1] Al-Jahiz was covered as a zoologist in Volume 1 for the Natural Scientists; though he also was a medical scientist and could have been covered in this volume for the Medical Scientists.

Kamil described a regular pentagon using a 4th order equation, thus laying down the foundation for analytic geometry.

6. Ibn Yunus' contributed methods for determining the time from solar or stellar altitude. Two of these methods together were equivalent to the trigonometric identity

    2 cos(a)cos(b) = cos(a+b) + cos(a-b)

    which is attributed entirely to Johannes Werner.

7. Ibn Al-Haytham wrote his seven-volume monumental treatise titled كتاب المناظر, "Book of Optics" which was thitherto the only treatment of optics that could genuinely be described as scientific. In comparison, the Greek treatment was pedestrian and erroneous; who, including Euclid and Ptolemy, believed in an "Emission Theory" whereby the eye is the emitter of light. Without the work of Ibn al-Haytham, the Six Books of Optics by Franciscus Aguilonius, works by Christiaan Huygens, and Hering's law of equal innervation, could not have been possible[2].

8. Ali Ibn Khalaf and Al-Zarqali invented the Universal Astrolabe and its design was copied in the Libros del Saber of Alfonso X of Castille in Spain.

---

[2] Ibn al-Haytham was covered as a mathematician, astronomer, and physicist in Volume 1 for Natural Scientist. He also worked on the functioning of the human eye, which falls under medical sciences. The law proposes that conjugacy of saccades is due to innate connections in which the eye muscles responsible for each eye's movements are innervated equally.

9. Yusuf al-Mu'taman ibn Hud wrote the monumental treatise titled "Kitab al-Istikmal" in which he proved what is now called Ceva's Theorem.

10. Ali Ibn Ridwan astronomically observed the brightest astronomical event in recorded history, the Supernova SN1006. Ibn Sina also observed it in Iran. Some others observed it with naked eye without collecting astronomical data. SN1006 is not attributed to Ali Ibn Ridwan, though such observations are often attributed to the Astronomer who observed them, like SN 1987A is attributed to Canadian astronomer Ian K. Shelton.

11. Al-Jayyānī researched the Unknown Arcs of a Sphere and wrote a comprehensive treatise on Spherical Trigonometry, presenting formulae for right-handed triangles, the general law of sines, and the solution of a spherical triangle by means of the polar triangle. These and his definition of ratios as numbers and his method of solving a spherical triangle when all sides are unknown are aspects that Regiomontanus later reported in his own works.

12. Al-Zarqālī noted that the path of the center of the primary epicycle is not a circle, as it is for the other planets. Instead, it is approximately oval contributing the existence of non-circular orbits, such as elliptic orbits.

13. Omar Khayyam solved the cubic equations in mathematics by connecting the cubic equation with geometric conic sections, thus further laying a firm foundation for analytic geometry. He provided proofs in algebra, presented methods for extracting n^{th} root of a number, and researched the Binomial theorem. One of Khayyam's predecessors, Al-Karaji, had already discovered the triangular arrangement of the coefficients of binomial expansions, and Khayyam popularized this triangular array in Iran, so that it is known as Omar Khayyam's triangle. However, later it came to be called as Pascal's triangle. Omar Khayyam proved a connection of Euclid's parallel axiom with the 4th postulate, in an elaborate attempt to prove the parallel axiom itself. This proof clearly showed the possibility of non-Euclidean geometries. However, these geometries now carry the names of Riemann and Gauss, Bolyai, and Lobachevsky.

14. Jabir Ibn Aflah corrected Almagest of Ptolemy in In his book "Iṣlāḥ al-Majisṭi". However, Regiomontanus reported the results in his own book, named "On Triangles".

15. Nur ad-Din al-Bitruji proposed a physical cause of celestial motions in his book of theoretical astronomy and cosmology, (كتاب الهيئة), which was an explicit search for a force like gravity to drive the astronomical bodies in their orbits. Some writer

used the material in a treatise on tides, Escorial MS 1636, dated 1192.

16. Ismail Al-Jazari invented the foundational elements in mechanical engineering, and invented clocks and machines that were advanced in automation and robotics. "The Book of Knowledge of Ingenious Mechanical Devices" (Arabic: كتاب في معرفة الحيل الهندسية) is a detailed analysis of 50 machines and how to construct them using fundamentally new concepts in mechanical engineering, like camshaft, crankshaft, segmental gear, control design, pump with valves, reciprocating piston motion, and various water raising mechanisms. He used hydropower for automation, a technique later used by Leonardo da Vinci. Al-Jazari invented robots for a waitress's work, hand-washing mechanisms, peacock fountains with servants, robotic music band, water clock with drummers with robotic servants, and castle clocks.

17. Al-Urdi wrote Kitāb al-Hay'a (كتاب الهيئة), a work on theoretical astronomy, in which he proved a mathematical lemma (Urdi Lemma) to generalize Apollonius theorem to allow an equant in an astronomic model to be replaced with an equivalent epicycle that moved around a deferent centered at half the distance to the equant point: a result upon which Copernicus later based his work.

The work of Nicolaus Copernicus has uncanny similarities between his work and the uncited work of Muslim astronomers, including, specifically, Nasir al-Din al-Tusi, Ibn al-Shatir, Muayyad al-Din al-Urdi, and Qutb al-Din al-Shirazi. There are unmistakable similarities in the Tusi couple and Copernicus' geometric method of removing the Equant from mathematical astronomy. *Not only do both of the methods match geometrically, more importantly they both use the same exact lettering system for each vertex; a detail that seems too preternatural.* Moreover, the fact that several other details of his model also mirror the work of other Muslim astronomers, begs the question if Copernicus' work was his own.

Muslim scientists are the giants on whose shoulders present-day sciences are built. All the Muslim scientists described in this book are of a giant stature; however, the book includes but only a few from the Muslim science community in Islamic Era. There are innumerable others, and many have been lost to oblivion. There is a wealth of "science" buried in that community and it remains to be extracted from the archives. Researchers will no doubt make further discoveries. Subsequent editions of this book would expand the set of scientists included, as well as additional details about those already covered.

There is at least a three-fold purpose to this book. One is to invite the world science community, in a manner of civilizational dialogue, to celebrate the science giants that Islamic Era has contributed to the growth of science and technology at its foundational level as well as at the level of expanding its frontiers. Another is to remind the Muslims of their love for "science" which every man and woman must acquire; not for worldly dominance, but for a better humanity in a better world. One other objective is to join hands with the rest of humanity by satisfying the upwelling desire of the youth to know the truth about Muslim civilization and the excellence of their pursuit for wholistic knowledge: scientific, humanitarian, cultural, civilizational, and spiritual.

It is time for the world to move ahead of the historical biases, religious prejudices, cultural entanglements, and hegemonic aspirations. All people, together, constitute our humanity, and we hold this truth as self-evident that all humans are created with equal value. So, let us all join hands to work together to make science in the service of making every day a wonderful day in every neighborhood of our planet.